About Island Press

Since 1984, the nonprofit organization Island Press has been stimulating, shaping, and communicating ideas that are essential for solving environmental problems worldwide. With more than 1,000 titles in print and some 30 new releases each year, we are the nation's leading publisher on environmental issues. We identify innovative thinkers and emerging trends in the environmental field. We work with world-renowned experts and authors to develop cross-disciplinary solutions to environmental challenges.

Island Press designs and executes educational campaigns, in conjunction with our authors, to communicate their critical messages in print, in person, and online using the latest technologies, innovative programs, and the media. Our goal is to reach targeted audiences—scientists, policy makers, environmental advocates, urban planners, the media, and concerned citizens—with information that can be used to create the framework for long-term ecological health and human well-being.

Island Press gratefully acknowledges major support from The Bobolink Foundation, Caldera Foundation, The Curtis and Edith Munson Foundation, The Forrest C. and Frances H. Lattner Foundation, The JPB Foundation, The Kresge Foundation, The Summit Charitable Foundation, Inc., and many other generous organizations and individuals.

The opinions expressed in this book are those of the author(s) and do not necessarily reflect the views of our supporters.

PRIMER OF ECOLOGICAL RESTORATION

PRIMER OF ECOLOGICAL RESTORATION

Karen D. Holl

Published in cooperation with the
Society for Ecological Restoration

 ISLANDPRESS

Washington | Covelo | London

Library of Congress Control Number: 2019948297

All Island Press books are printed on environmentally responsible materials.

Manufactured in the United States of America
10 9 8 7 6 5 4 3 2

All restoration project costs in this book are listed in US dollars. Measurements are given in metric units. To convert to US equivalents:
1 millimeter = 0.04 inch
1 centimeter = 0.39 inch
1 meter = 3.28 feet
1 kilometer = 0.62 mile
1 hectare = 2.47 acres

Keywords: adaptive management, aquatic, ecosystem, exotic species, fauna, hydrology, invasive species, landform, legislation, monitoring, nonnative species, paying, planning, reclamation, rehabilitation, restoration ecology, restore, revegetation, river, soil, terrestrial, vegetation, water

Contents

To Travis and my students, with hopes for a more sustainable future

Preface

The science and practice of ecological restoration have grown exponentially over the past few decades as we aim to compensate for the negative impacts humans have had on the ecosystems on which we and millions of other species depend. Increasingly, we undertake restoration because of its importance for ensuring our own well-being. With the growth of ecological restoration has come a plethora of resources: thousands of articles in the peer-reviewed and management literature, countless websites describing individual projects, a few general textbooks, and many books focused on restoring specific ecosystem types. The information I review in *Primer of Ecological Restoration* is not new. Rather, my goal is to provide a broad but succinct introduction to ecological restoration for a few audiences. First, I anticipate that this book will be used as an introductory text for some ecological restoration and restoration ecology courses in which the instructors assign students in-depth readings on specific topics and case studies tailored to the focus of the course. Second, this book could be used as one of a few texts in courses on conservation biology and resource management where ecological restoration is not the only topic covered. Third, this book should be of interest to natural resource managers and a more general audience who want a short introduction to ecological restoration. To that end, I have kept specialized jargon to a minimum and define terms in both the text and the glossary.

Restoring ecosystems requires an interdisciplinary background. Understanding the ecology and natural history of the ecosystem being restored and knowing appropriate restoration methods are essential. But, as any restoration practitioner knows, successful project implementation requires familiarity with a host of other topics, including but not limited to managing stakeholder involvement and public outreach; experience with planning, goal setting, and monitoring; and knowledge of relevant legislation, permitting processes, and funding sources. This book could not possibly discuss all these topics in detail while achieving the goal of brevity, so instead I provide an overview of key points and illustrate them with brief examples. These different topics are necessarily in separate chapters in this book, but they must be synthesized when designing and implementing a restoration project. Concepts are integrated here by cross-references to chapters and several online case studies that provide detailed information and that integrate various themes illustrated by the project.

The old saying that a picture is worth a thousand words is true for ecological restoration. There is no substitute for seeing before and after photos of projects and visiting restoration sites in person. Nonetheless, because including numerous color photos in the book would have greatly increased the cost and hence made the book less accessible to a broad audience, I chose instead to use selected diagrams and tables in the book and to incorporate photos in the online case studies. Listed on the book's website (islandpress.org/restoration-primer) are links to a few of the many restoration project websites, photos, and videos available on the internet, and I encourage you to visit restoration projects in your area.

This book is not intended as a thorough guide of how to restore specific ecosystem types, so readers who plan to work in the field of ecological restoration will want more in-depth resources on specific topics. To this end, I have provided short reading lists at the end of each chapter. On the website, I also provide questions for reflection and discussion that ask readers to apply the ideas presented in the book to a restoration project of their choice.

As with any book, the content reflects the biases and experiences of the author. Although I am trained as an ecologist, I have worked closely with political scientists, economists, and natural resource managers over my career. I am a professor at a research university, where I have taught an interdisciplinary course on restoration ecology and researched ecological questions underlying restoration for more than two decades. I have advised, but do not have extensive experience implementing, large restoration projects. I have worked in a range of terrestrial ecosystems in several regions, including hardwood forests in the eastern United States; grassland, chaparral, and riparian forest ecosystems in California; and tropical rain forests in various countries in Latin America. My knowledge of restoration lake and wetland ecosystems is primarily from the work of others. I am committed to bridging the divide between academic and management communities so that scientific research is best designed to inform and improve on-the-ground restoration projects, a commitment that is reflected in the contents of this book. Finally, I am passionate about educating the next generation of restoration ecologists, a goal I hope this book will help fulfill.

Acknowledgments

I owe a huge debt of gratitude to many people who helped make this *Primer of Ecological Restoration* a reality. Many thanks to Josie Lesage, my right-hand woman on this project, who helped with numerous aspects of this book over the last two years; she edited most chapters more than once, filled in examples, wrote a case study, and gave me extensive feedback throughout the process. I also appreciate Alicia Calle, Michelle Pastor, and Michael Baca for contributing their artistic skills to various figures.

Pedro Brancalion, Virginia Matzek, Leighton Reid, and Jose Maria Rey Benayas undertook the large task of reviewing the entire book manuscript, and James Aronson read several chapters. I am grateful for their candid and insightful feedback, which improved the book immensely. A special thanks to Leighton Reid, whom I frequently asked for feedback and who always replied promptly with thoughtful answers. Many people contributed their expertise to specific chapters, including Alicia Calle, George Gann, Brent Haddad, Andy Kulikowski, Michelle McCrackin, Tein McDonald, Adam Millard-Ball, Margaret Palmer, Daniel Press, Alexandre Sampaio, Isabel Schmidt, Drew Scott, Rachel Shellabarger, Joy Zedler, and Margaret Zimmer. If I did not address all your comments, it was due to length constraints, and any errors that remain are mine. I am also grateful to the people who coauthored or gave feedback on the online case studies: Peter Brewitt, Ben Brown, Tom Dudley, Amy East, James Gibbs, Greg Golet, Beth Howard, Joe Koebel, Ryan Luster, K. A. Sunanda Kodikara, Joe Silveira, and others who are mentioned above.

I appreciate feedback from many students in the 2018 and 2019 Restoration Ecology classes at the University of California, Santa Cruz who read some or all of the draft and provided helpful input on points that needed clarification; Joia Fishman, Lexi Necarsulmer, and Emily Reynolds in particular deserve recognition for their detailed comments.

Erin Johnson at Island Press has provided ongoing encouragement and patient feedback on my many queries throughout the book writing process. Thank you.

Thank you to my parents, who have supported me at every stage of my education and career. Most of all, to Michael and Travis go my thanks for listening to me discuss this book countless nights at dinner. Michael, I appreciate your patiently enduring discussions of various versions of this book project for many years. And Travis, I hope this book will have a positive benefit on the environment for your generation.

1

Why Restore Ecosystems?

The enormous extent of human impacts on Earth has caused many to propose that we are now in the age of the "Anthropocene," a human-dominated geological epoch (Crutzen 2002). Humans have influenced *ecosystems*[1] for thousands of years in many ways, from managing species of agricultural value and altering water flow patterns to irrigate crops to using fire as a tool to clear lands and increase soil fertility. Regardless of previous impacts, the pace, intensity, and scale at which humans have altered the planet in recent decades are unprecedented. At this point, even the most remote locations on Earth have been influenced by anthropogenic climate change and long-distance transport of pollutants, and less than one-fourth of the land area is free of direct human impact (Intergovernmental Science-Policy Platform on Biodiversity and Ecosystem Services 2018).

These impacts come in many forms. Staggering figures can be cited for the loss of all types of ecosystems in every region worldwide. Human activities have resulted in the destruction of more than 10 percent of the dwindling wilderness in the world between the early 1990s and 2015 (Watson et al. 2016), the transformation of 38 percent of global land area for agriculture (FAO n.d.), and the *degradation* of many remaining ecosystems by human activities such as logging, overhunting, mining, and fire suppression. Anthropogenic changes to hydrologic patterns have dramatically transformed most rivers and wetlands. Human activities have substantially

---

1. Italicized terms are defined in the glossary.

increased levels of phosphorus and biologically available nitrogen and have resulted in toxic concentrations of many substances in the air and water. In addition to local and regional impacts, human activities are rapidly increasing concentrations of greenhouses gases in the atmosphere. These gases cause changes in global climate patterns, including increased temperature, altered precipitation, rising sea levels, and an increasing frequency of extreme weather events. Elevated carbon dioxide levels also directly affect plant growth and drive acidification of the oceans.

These local, regional, and global transformations of ecosystems jeopardize human well-being in numerous ways (Potts et al. 2018). Land degradation has direct impacts on public health because the loss of forest, grassland, and wetland ecosystems that filter pollutants from water results in an increasing number of people who do not have access to safe drinking water. Destruction of coastal ecosystems elevates the risk of shoreline inhabitants to increasingly frequent and intense storms and causes increased migration. Land degradation costs the world an estimated $6.3 trillion to $10.6 trillion per year, equivalent to 10 to 17 percent of the global gross domestic product (ELD Initiative 2015). Equally noteworthy is that ecosystem degradation exacerbates income inequity; the rural poor obtain a larger share of their income directly from noncultivated resources, such as firewood, construction materials, fisheries, and other food products, so they feel the effects disproportionately (Potts et al. 2018).

Conserving species, ecosystems, and, ultimately, humans, will require dramatic changes to resource distribution and consumption patterns, as well as slowing the human population growth rate. These vast and pressing topics have been discussed extensively elsewhere. One important complementary strategy to counteract the extensive human impacts on the natural world is to restore damaged ecosystems.

Motivations for Restoration

The term *ecological restoration* is used in different ways (chap. 2) but most commonly is defined as the "process of assisting the *recovery* of an ecosystem that has been degraded, damaged, or destroyed" (Society for Ecological Restoration Science and Policy Working Group 2004). Ecological restoration is driven by a diverse and overlapping set of reasons (table 1.1; Clewell and Aronson 2006, 2013).

Most ecological restoration projects are motivated, at least in part, by a desire to bring back species, ecosystems, or *ecosystem processes* (e.g., nutrient *cycling*, primary *productivity*, seed dispersal) that have been compromised by human activities. Increasingly, restoration projects are prompted

Table 1.1. Motivations for Restoring Damaged Ecosystems

Category	Motivation	Examples
Compensation for past damages	Conserving *biodiversity*	Species and *habitat* types
	Enhancing *ecosystem processes*	Primary *production*, nutrient cycling
	Counteracting climate change	*Carbon storage*, coastal erosion
Economic	Providing *ecosystem services*	Water purification, crop pollination, timber and nontimber forest products
	Providing employment and job training	Stream restoration and *invasive* species removal
Cultural/ spiritual	Reconnecting with nature and experiential education	Local adopt-an-ecosystem projects
	Conserving cultural values	Restoring species important to cultural heritage
	Atoning for past damages	Personal renewal through participating in volunteer restoration projects
Legislative	Complying with legislation	Various laws requiring restoration (e.g., wetland mitigation, mine reclamation, restoration of habitat of *endangered species*)

by an attempt to mitigate for and adapt to climate change. Forest, wetland, and grassland restoration can increase *carbon storage* and, along with aggressive efforts to reduce carbon emissions, can help reduce global temperature increase. Restoration of coastal ecosystems, such as mangroves and coral reefs, is a cost-effective way to reduce risks from storms (Asian Mangrove case study[2]). Ecological restoration can help humans and ecosystems adapt to climate change in various ways, such as providing *refugia* for climate-sensitive species and improving the resilience of crop production to climate variability (Locatelli et al. 2015).

Ecological restoration provides extensive economic benefits to humans through *ecosystem services*, which are the suite of benefits that ecosystems provide to humanity (Millenium Ecosystem Assessment 2015). They range from supplying people with food, medicines, and fuel to providing important functions such as water purification, flood control, and crop pollination. They are goods and services that the natural world has always

---

2. Case studies are available on this book's website: islandpress.org/restoration -primer.

provided to humans, but that we have frequently overlooked until after ecosystems are destroyed or degraded. Restoring an ecosystem is often a less expensive option to provide humanity with specific services than trying to provide the service with a heavily engineered solution. For example, Ferrario et al. (2014) found that on average the cost of installing seawalls and breakwaters was at least ten times more expensive than restoring reefs to provide storm protection to coastal cities. Moreover, some ecosystem services are simply irreplaceable at any cost. Whereas engineered structures may substitute for the coastal erosion control services that reefs provide, they do not provide the recreational values of people who visit reefs, cultural values of indigenous groups that have relied on reefs for fisheries for generations, and the *biodiversity* hosted in reefs that might provide compounds for pharmaceuticals.

Some restoration projects are funded as job creation and training programs to provide direct economic benefits. For instance, the Working for Water Project in South Africa has employed approximately 10,000 workers yearly between 1996 and 2012 on projects to remove invasive nonnative trees and shrubs that reduce water supply; this program has been driven largely by the government's aim to increase rural employment (van Wilgen and Wannenburgh 2016).

A host of social and cultural factors also motivate restoration projects (Clewell and Aronson 2006; Egan, Hjerpe, and Abrams 2011). Many restoration projects are led by individuals and community groups who want to restore a local ecosystem because of a sense of connection to the land or for aesthetic reasons (Dolan, Harris, and Adler 2015). Restoration can provide opportunities for place-based education for learners of all ages; as such, a growing number of restoration curricula are available that integrate science standards with hands-on restoration experiments. Such local projects may offer participants an opportunity for spiritual renewal and to atone for damages caused by humans (Jordan 2003). Likewise, some projects, particularly those involving indigenous groups, focus on restoring cultural values, such as replanting or managing for certain traditional plants used by indigenous groups for food or basketry (Uprety et al. 2012).

The need to compensate for past human damage to ecosystems—combined with the ecological, economic, and social benefits of restoration—has resulted in a host of legislative mandates to fund or require restoration (chap. 11). In many countries, laws require mining companies to restore or *reclaim* lands after mining is completed. In some countries, laws target restoration of specific ecosystems, such as wetlands in the United States.

Together, these motivations have resulted in calls for the restoration of hundreds of millions of hectares of land at the global scale. It is important to recognize that even within individual restoration projects, different people and organizations are motivated to restore for different reasons. Hence, discussing and aiming to meet different *goals* and desires is a critical part of the planning process (chap. 3; Gann et al. 2019).

Restoration as One Component of Conservation Efforts

Ecological restoration is one of a suite of strategies to conserve biodiversity, ecosystems, and the services these ecosystems provide to humans. Clearly, protecting and maintaining minimally impacted ecosystems should remain at the core of conservation practice given that many research projects and case studies show that even the most successful restoration projects restore a subset of the species and ecosystem services present prior to *disturbance* (Rey Benayas et al. 2009; Moreno-Mateos et al. 2017).

Academics once debated whether humans should intervene to help facilitate the recovery of damaged ecosystems or just allow the ecosystems to recover on their own. Today it is widely recognized that human *management* to restore ecosystems is an important complementary component of conservation efforts given the intensity and extent of existing human impacts and the need to replace lost ecosystem services to people as quickly as possible. The question is no longer whether to restore ecosystems, but, rather, in what cases and to what extent should we intervene to facilitate ecosystem recovery? In addition, when should we prioritize restoration among the range of conservation actions?

Restoration efforts have been criticized for undermining *habitat* preservation efforts by offering an opportunity to offset habitat destruction, yet I contend that few, if any, restoration ecologists would suggest restoration as an alternative to habitat preservation. When a person's house is burglarized, a primary concern is to improve security so that the act is not repeated, but improving security does not lessen the need to replace stolen items. Of course, there may be no substitutes for certain items such as photographs or other memorabilia, but the owners normally do their best to re-create the house as it was before the vandalism. Likewise, conservation and restoration are not mutually exclusive; they are complementary actions. In general, the field of conservation biology has become more hands-on in recent years (Hobbs et al. 2011); actions are increasingly taken to maintain existing habitats both proactively (e.g., preventing invasive species from colonizing existing habitats) and reactively (e.g., removing invasive species).

Whereas restoration may mitigate some anthropogenic impacts on the natural world, restoration is a useless exercise unless it is part of an effort to reduce the drivers of habitat conversion, which are complex and vary across the globe (Geist et al. 2006). The human population continues to grow rapidly, having increased by 1.6 billion people between 2000 and 2019, and we are adding approximately 200,000 additional people to the planet each day. Likewise, high levels of consumption in places like the United States and Europe and growing levels of consumption in nations such as China and Brazil increase human impacts on ecosystems. Complex patterns of global trade and rural-urban migration, as well as new technologies, interact to affect land use patterns (Lambin and Meyfroidt 2011). Although a detailed discussion of how to reduce these drivers of habitat degradation and conversion is beyond the scope of the book, it is critical to recognize that ecological restoration has to be a part of multifaceted efforts to conserve ecosystems while providing for human *livelihoods*. A broad range of approaches is needed not only to conserve and restore ecosystems and species, but also to ensure the survival of the human species that depends on them.

Recommended Reading

Clewell, Andre, and James Aronson. 2006. "Motivations for the restoration of ecosystems." *Conservation Biology* 20:420–28.
>   Discusses a range of motivations for restoring damaged ecosystems.

Egan, David, Evan E. Hjerpe, and Jesse Abrams (eds). 2011. *Human Dimensions of Ecological Restoration*. Washington, DC: Island Press.
>   Each chapter of this edited volume addresses different aspects and case studies of human participation in restoration projects.

Intergovernmental Science-Policy Platform on Biodiversity and Ecosystem Services. 2018. *Summary for Policymakers of the Assessment Report on Land Degradation and Restoration of the Intergovernmental Science-Policy Platform on Biodiversity and Ecosystem Services.* Bonn, Germany: IPBES Secretariat. https://www.ipbes.net/assessment-reports/ldr.
>   Summarizes an international effort to provide a current review of the state of the knowledge of land degradation and restoration.

2

Defining Restoration

Given the many different motivations for restoration (chap. 1) and the broad range of strategies used to restore *ecosystems*, it is not surprising that definitions of restoration are also broad and variable. In the early years of the field of restoration ecology, there was a strong distinction between the term *restoration* and other terms describing ecosystem *management* with different goals. *Restoration* was used to refer to efforts to restore "predisturbance" or "historical" *community composition, ecosystem structure*, and *ecosystem processes* (fig. 2.1). In contrast, other terms, such as *rehabilitation, reclamation*, and *revegetation*, described efforts to improve the condition of a degraded ecosystem, typically focusing on specific ecosystem processes and *ecosystem services*, such as enhancing plant productivity, reducing erosion, or improving water quality, without necessarily striving to re-create a specific community composition (table 2.1; Bradshaw 1984).

Over time, the definition of restoration has continued to evolve and be the subject of extensive debate. The most commonly used definition of restoration in the literature is from the Society for Ecological Restoration (Society for Ecological Restoration Science and Policy Working Group [SER] 2004, 3), namely that *ecological restoration* "is the process of assisting the recovery of an ecosystem that has been degraded, damaged, or destroyed." Under this definition, the general target of restoration is "a characteristic assemblage of the species that occur in the reference ecosystem and that provide appropriate community structure." The aim is still to set the ecosystem on a *trajectory* toward recovering community composition,

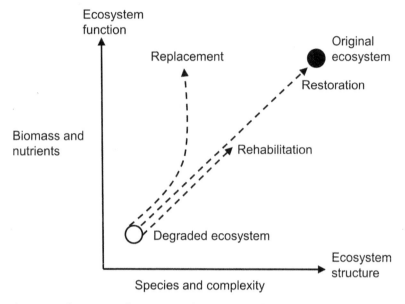

Figure 2.1. Classic view of restoration where anticipated recovery follows a linear recovery path toward the original ecosystem. Figure simplified and redrawn from Bradshaw 1984.

ecosystem structure, and ecosystem processes within the historical range of variability, but there is increasing recognition that even minimally disturbed ecosystems are variable over space and time, so there is not a single endpoint (fig. 2.2; SER 2004; Palmer, Falk, and Zedler 2006). The 2019 SER International Standards for the Practice of Ecological Restoration (Gann et al. 2019), a recent attempt to standardize the terminology, principles, and practices of restoration, defines full *recovery* as the "state or condition whereby all the key ecosystem attributes closely resemble those of the reference model," recognizing that, in fact, there is a range of states within the natural variability rather than a single static endpoint.

Suding et al. (2015) suggest that rather than using a single definition, restoration projects should be evaluated based on four principles, namely whether restoration (1) increases ecological integrity, (2) is sustainable in the long term (i.e., does not require ongoing human intervention), (3) is informed by the past and future, and (4) benefits and engages society. They assert that if a project meets all these principles, then it is "restoration"; if it does not, it may fall into one of the other definitions, such as *compensatory mitigation* or rehabilitation of *ecosystem services*. Gann et al. (2019) state that the ecological restoration practice should (1) be effective

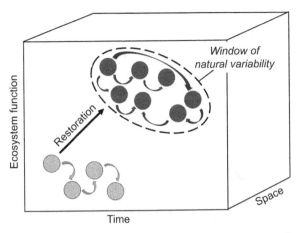

Figure 2.2. A more dynamic view of ecosystem recovery. All attributes of natural or reference ecosystems (dark gray spheres) vary over time and space within a natural window of variability (dashed oval line). When ecosystems move outside that window of natural variability (light gray spheres), restoration aims to set the ecosystem on a trajectory toward the natural or historic window of variability. Figure redrawn from Palmer, Falk, and Zedler 2006.

in restoring ecosystem "values," (2) maximize beneficial outcomes while minimizing resource inputs, and (3) engage partners and *stakeholders* (chap. 3). They frame a "family of restorative activities" that, like earlier terms, varies along a continuum in the degree to which it aims to restore different ecosystem attributes.

There is extensive discussion about the terms used to distinguish the degree of human intervention in the recovery process. *Natural regeneration*, passive restoration, and spontaneous regeneration refer to removing the degrading factors, such as agriculture or grazing, and allowing the ecosystem to recover through natural colonization of plants and animals rather than actively reintroducing species (Holl and Aide 2011; Gann et al. 2019). *Assisted regeneration* is an intermediate restoration approach to facilitate the recovery process in sites that show some natural regeneration by taking actions such as removing pest organisms or reintroducing ecological disturbance regimes (Gann et al. 2019). *Active restoration* or reconstruction (a term used by Gann et al. 2019) refers to a restoration approach whereby there is extensive human intervention to accelerate recovery, often by reintroducing many or all of the species.

At first glance, the difference between these activities seems clear, but deciding where specific restoration activities fall along the continuum can

Table 2.1. Definitions of Restoration and Related Activities

Term	Definition	Reference
Active restoration, Reconstruction	A restoration approach in which there is extensive human intervention to influence the rate of recovery and the arrival of the biota is largely or entirely dependent on human agency.	Holl and Aide 2011; Gann et al. 2019
Assisted regeneration	A restoration approach that focuses on actively harnessing any natural regeneration capacity of biota remaining on-site or nearby.	Gann et al. 2019
Ecological restoration	Re-creating functioning ecosystems in which plants, animals, and soil are functioning within the normal range of activity. Aiming for a close resemblance to what was there before.	Bradshaw 1984
Ecological restoration	The process of assisting the recovery of an ecosystem that has been degraded, damaged, or destroyed.	SER 2004
Forest and landscape restoration	A planned process that aims to regain ecological functionality and enhance human well-being in deforested or degraded landscapes.	Reitbergen-McCracken, Maginnis, and Sarre 2007
Mitigation	A series of actions taken to minimize the environmental damage of a development or danger to a *species of concern*. Steps include avoiding project alternatives that would be particularly damaging, modifying the project to minimize negative impacts to the degree possible, and compensating impacts that cannot be avoided through *compensatory mitigation*.	Gann et al. 2019
Natural regeneration, Spontaneous regeneration, Passive restoration	An approach to restoration that relies on spontaneous increases in biota without direct reintroduction after the removal of degrading factors alone.	Holl and Aide 2011; Gann et al. 2019
Reclamation	Making severely degraded land (e.g., former mine sites or wastelands) fit for cultivation or a state suitable for some human use. Emphasis is on returning the site to an anthropocentrically useful condition or trajectory.	Bradshaw 1984; Gann et al. 2019
Reforestation	Planting trees on lands that were previously forested. The species used may or may not be native.	Gann et al. 2019
Rehabilitation	Actions that aim to reinstate a level of ecosystem functionality where ecological restoration is not sought, but rather the focus is on the provision of ecosystem goods and services.	SER 2004; Gann et al. 2019
Revegetation	Establishment, by any means, of plants on sites that may or may not involve local or *native species*.	Gann et al. 2019
Rewilding	Restoring an area of land to its uncultivated or "wild" state. Used with reference to the reintroduction of wild animals that have been exterminated, as well as restoring other ecological processes.	Corlett 2016; Perino et al. 2019

be complicated (Gann et al. 2019). Is removing a dam to allow for *riparian* forest recovery natural regeneration, assisted regeneration, or active restoration? It removes the human impact of restricting water flow without actively reintroducing species, but requires considerable human intervention and resources. Authors disagree over whether removing pest species is removing a degrading factor to allow for natural regeneration or whether it is assisted regeneration. Clearly, the degree of intervention varies along a gradient rather than there being distinct categorical differences, and what constitutes natural regeneration versus assisted regeneration versus active restoration varies by ecosystem and disturbance types. The choice of restoration approach depends in large part on the recovery rate of the ecosystem and the specific aims of the project (Holl and Aide 2011; Gann et al. 2019). For example, extensive human intervention is most appropriate in highly degraded systems where natural regeneration would not occur within the desired time period.

Why Is Restoring to a Historical State So Difficult?

The array of restoration definitions discussed above reflects that defining the historical state or range of variability is challenging, and achieving that state is even more difficult, for several reasons discussed below. Moreover, restoration is undertaken for a host of different reasons in our rapidly changing world (chap. 1).

Defining the Historical Target and Shifting Baselines

Restoration projects often aim to restore an ecosystem to a state prior to human *disturbance*. However, the question of what past temporal state to select is subjective (Aronson, Dhillion, and Le Floch 1995; Barak et al. 2016). Do we restore to fifty, two hundred, or one thousand years ago? In the Americas, it is common to aim to restore ecosystems to a state prior to European colonization, but that is an arbitrary point in time given that indigenous peoples influenced these landscapes for thousands of years. The question is likewise complicated in Europe, where records indicate that humans have actively manipulated ecosystems extensively for thousands of years (Backstrom et al. 2018). In these regions, *traditional cultural ecosystems*, which have developed with historical human land use practices, may help inform the *reference model* (Gann et al. 2019). In addition, ecosystems are not static and have changed in response to nonhuman-caused fluctuations in climate and weather on the time period of decades, centuries, or longer (Millar and Brubaker 2006). The specific time point selected strongly affects the restoration target.

If the goal of restoration is a historical state, then a common confounding problem is the lack of detailed information about what the ecosystem looked like. If the reference model (chap. 3) is what the site looked like a decade or two ago, then detailed information on species composition and ecosystem processes should be readily available. In contrast, if a restoration project in the Americas aims to restore to a state prior to European colonization, then knowledge on the species composition at that time relies largely on limited natural history notes of the early European explorers and, in rare cases, on ethnographic accounts and drawings from indigenous peoples. Characterizing the composition of historical ecosystems in regions with long periods of extensive agricultural use is nearly impossible. At best, one can gain a general vision of what the ecosystem looked like rather than the details needed to set specific restoration objectives.

A related issue is that what people perceive as a "historical" or "pre-disturbance" state is subject to human interpretation (Backstrom et al. 2018). With the unprecedented changes in the scale of human impacts on ecosystems over the past few decades (chap. 1), it is increasingly apparent that even ecosystems considered as "pristine" or "wilderness" are changing in response to anthropogenic impacts that often occur due to actions far from a given site. As that happens, we become more accustomed to the altered state, a phenomenon known as "shifting baselines" (Pauly 1995); in other words, each successive generation of people assumes that the diminished biological state is the norm rather than recognizing that this state has itself been altered by prior human activities. Pauly originally described shifting baselines in the context of fisheries where scientists compare fish declines to the abundance at the start of their careers without considering historical declines in fish populations due to overfishing over a period of centuries. One can think of numerous examples where this perception is the case, such as historical extinction of many faunal species due to forest clearing and overhunting centuries ago that have in turn affected the dispersal and distribution of plant species more recently. These changes mean that restoration *practitioners* in each successive generation are likely to lower their expectations for ecosystem recovery. Shifting baselines also make it difficult to judge whether a restored system is returning to the reference model (chap. 3) if the reference ecosystem is changing at the same time.

Impossibility of Controlling the Trajectory of Ecosystem Recovery

Once a reference model for restoration is chosen, another challenge is directing the trajectory of recovery toward the desired state. Even in minimally human-impacted systems, ecosystem recovery is often highly variable and unpredictable (chap. 5) as opposed to the linear trajectory of recovery that is often assumed (see fig. 2.1). Interannual weather fluctuations, natural disturbances (e.g., fire and flooding), and rare long-distance dispersal events affect the species that establish. For example, the species that colonize and survive in the first few years of dry forest restoration are determined by which plants set seed, the amount of rainfall in each year, whether a fire burns through the site, and an infinite number of other factors. The ecosystem trajectory in subsequent years is affected by which species establish initially, as well as ongoing variation in climatic variables, natural disturbances, and other stochastic events. The result is what Hilderbrand, Watts, and Randle (2005) refer to as the "myth of the carbon copy" in restoration. In other words, we can invest considerable resources to restore *abiotic* conditions and introduce desired species, but it is impossible to control the many variables at multiple spatial and temporal scales that influence the trajectory of ecosystem recovery and re-create a prior ecosystem exactly.

Lack of Resources and Knowledge

Billions of dollars are spent globally on restoration each year, but there will never be sufficient funds to undertake all the necessary restoration work. Limited funding makes it difficult to support projects over the multiple years necessary to ensure that the resulting ecosystems resemble the historical state that was selected. There are a number of high-profile and well-funded projects, such as restoration of 100 kilometers of the Kissimmee River in Florida, which has cost approximately $800 million (Kissimmee River case study). Most restoration projects, however, are undertaken with limited funds and are often supported by volunteer labor. There will always be trade-offs between what is desirable and what is feasible given funding availability.

Likewise, lack of knowledge limits our ability to restore ecosystems fully. Our understanding of how to restore many ecosystems has advanced substantially through a mix of scientific studies and learning from our successes and failures in restoration projects. Nonetheless, major gaps remain in our knowledge of the complex interactions between abiotic and *biotic* factors in nearly all ecosystems, and we know even less about how to restore them. An apt analogy is Humpty Dumpty: like an egg, it is much easier to destroy an ecosystem than to put it back together again. The act of trying to restore ecosystems is the ultimate test of our understanding of how they work (Bradshaw 1987).

Conflicting Goals

On the surface, restoring an ecosystem to a specific historical trajectory seems relatively noncontroversial in terms of an ecological *goal*, but conflicts often arise regarding the desired ecosystem stage or *focal species*. As ecosystems recover, they go through a process of *succession* (chap. 5) during which disturbance-adapted species become less common and other species more abundant. Likewise, active restoration actions will favor some species over others. For example, as described in the Tamarisk Removal case study, removal of *invasive, nonnative* tamarisk (*Tamarix* spp.) trees to reduce transpiration of water and restore native riparian vegetation has been highly controversial because it negatively affects an endangered bird, the southwestern willow flycatcher (*Empidonax traillii extimus*), that nests in tamarisk trees, and conflicting ecological goals are only part of the challenge.

As the scale of both the human footprint on the landscape and the restoration projects being undertaken increases, balancing a host of ecological, socioeconomic, and cultural restoration goals is essential. This recognition is embedded in some recent definitions of restoration, which explicitly consider human well-being as a critical component of ecological restoration goals (Reitbergen-McCracken, Maginnis, and Sarre 2007; Suding et al. 2015). At the site scale, it might mean selecting nonnative tree species that are valued by local communities for fruit or timber as part of the planting palette for tropical forest restoration rather than only using native tree species. At the landscape scale, it means employing a mix of approaches to enhance *habitat* value for humans and other species. For example, *forest and landscape restoration* in tropical agriculture landscape mosaics often includes a mix of maintaining and restoring diversity in remnant forests, restoring low productivity agricultural lands that are important for minimizing erosion and flooding, and increasing and diversifying tree cover in actively used agricultural lands. These compromises between goals result in ecosystems that do not aim to replicate historical conditions across the entire landscape.

Novel Ecosystems Debate

In the twenty-first century, ecosystems are increasingly composed of non-historical or novel species assemblages (i.e., species combinations and relative abundances that have not been observed in recent human history) due to anthropogenic environmental changes, land conversion, species invasions and extinctions, or a combination of these factors (Hallett et al. 2013). For example, in California, nonnative eucalyptus trees now provide important overwintering habitat for native monarch butterflies (*Danaus plexippus*) whose populations are rapidly declining. These novel assemblages of species and shifting baselines of ecosystems further complicate the discussion of how to define and direct restoration efforts. Hobbs, Higgs, and Harris (2009) assert that restoration ecologists need to carefully consider *management* options and restoration targets depending on the extent of changes to a given ecosystem and whether the system has passed a threshold where restoring to a historical state would be extremely challenging and use extensive resources. In such cases, they argue that it would be wise to accept the target of a *novel ecosystem* that "retains characteristics of the historic system but whose composition or function now lies outside the historic range of variability" as a more reasonable goal (Hobbs, Higgs, and Harris 2009, 601). The authors contend that aiming for a novel state in some cases allows for more effective allocation of resources to other projects where restoration to a state within the historical range of variability is more likely.

This proposal to include novel ecosystems as a viable restoration target has met with strong opposition from some. Murcia et al. (2014) argue that this redefinition of restoration creates a great deal of ambiguity and provides justification for governments and others undertaking restoration to consider almost any ecosystem management as restoration, thus undermining recent large-scale commitments to restoration. They also argue that no clear thresholds exist for when an ecosystem can or cannot be restored to within the range of historical variability.

Higgs et al. (2014) provide a middle ground for the role of history in guiding restoration. They argue that we still need to consider past ecosystem composition in choosing a reference model, but that the role of historical states is shifting given the extent of environmental changes. They suggest that historical data can provide baseline information, tell us how ecosystems have changed over time, and serve as a guide for restoration efforts, but that it is often not realistic or pragmatic to aim to restore a copy of a prior state, particularly if it is not linked to human well-being. Many other authors (e.g., Suding et al. 2015; Falk 2017; Gann et al. 2019) concur that selecting a target for restoration will require a delicate balance between understanding the past while recognizing current and future conditions.

Definitions in Practice

The debate about appropriate definitions and broad goals for restoration continues, largely in the academic literature, with little likelihood of resolution. At the same time, policy makers and restoration practitioners continue to use the term *restoration* to refer to a range of goals and management actions. Efforts to clarify terms are valuable and important in a policy context, but completely standardizing the use of terms such as *ecological restoration* or *forest and landscape restoration* is impossible. Hence, as discussed in chapter 3, it is critical that those involved clearly define the goals and specific project *objectives* for each global restoration initiative and local restoration project. For example, is the goal of a specific forest restoration project to restore historical native plant cover, provide habitat for an endangered bird species, minimize erosion, provide income for landowners, sequester carbon, reduce source populations of invasive species, or more than one of those goals? Explicitly stating those goals is critical for transparency and honesty about the potential benefits and negative consequences of a project, for selecting the most appropriate restoration methods, and for evaluating whether the objectives have been achieved (chap. 3).

In addition, choosing a *reference model* that restoration aims to achieve is a subjective and value-laden decision (Backstrom et al. 2018). There is not one "clear" and "correct" reference model. Most restoration ecologists would agree that historical species composition and ecosystem functions should be considered when defining the reference model for a project, but specific ecological targets must be balanced with the many other considerations discussed above. What is certain is that different stakeholders will have varied opinions about restoration goals and the appropriate reference model, and these opinions should be discussed thoroughly at the outset if there is to be any hope of a clear vision and long-term success of restoration efforts.

Recommended Reading

Bradshaw, Anthony D. 1987. "Restoration: An acid test for ecology." In *Restoration Ecology*, edited by William R. Jordan III, Michael Gilpin, and John D. Aber, 23–29. Cambridge: Cambridge University Press.

Provides a historical perspective on the field of restoration ecology.

Gann, George D., Tein McDonald, Bethanie Walder, James Aronson, Cara R. Nelson, Justin Jonson, Cristina Eisenberg, et al. 2019. *International Principles and Standards for the Practice of Ecological Restoration.* Washington, DC: Society for Ecological Restoration.

Provides the most recent synthesis of terminology, principles, and standards of restoration that have been endorsed by the Society for Ecological Restoration, the professional society in the field of restoration ecology.

Higgs, Eric, Donald A. Falk, Anita Guerrini, Marcus Hall, Jim Harris, Richard J. Hobbs, Steven T. Jackson, et al. 2014. "The changing role of history in restoration ecology." *Frontiers in Ecology and the Environment* 12:499–506.

Presents a thoughtful discussion of how the role of history is changing in restoration ecology given ubiquitous anthropogenic impacts.

Hilderbrand, Robert H., Adam C. Watts, and April M. Randle. 2005. "The myths of restoration ecology." *Ecology and Society* 10:19.

Provides a summary and critique of five different assumptions on which most restoration ecology projects are based.

Hobbs, Richard J., Eric Higgs, and James A. Harris. 2009. "Novel ecosystems: Implications for conservation and restoration." *Trends in Ecology and Evolution* 24:599–605.

Introduces the novel ecosystems concept.

Murcia, Carolina, James Aronson, Gustavo H. Kattan, David Moreno-Mateos, Kingsley Dixon, and Daniel Simberloff. 2014. "A critique of the 'novel ecosystem' concept." *Trends in Ecology and Evolution* 29:548–53.

Describes the authors' concerns about the novel ecosystems concept.

3

Project Planning

Restoration projects come in all shapes and sizes. The people or groups who undertake projects do so for a variety of reasons, ranging from broad legislative requirements to a personal desire to contribute beneficially to society (chap. 1). Examples include government agencies with a mission to improve public lands, private companies complying with environmental legislation, and community groups seeking to enhance their local environment. Projects range in scale from a few square meters within a city block to thousands of hectares, such as the Kissimmee River restoration project in Florida (Kissimmee River case study).

Regardless of the size, complexity, or motivation behind a project, to be successful it must be carefully planned. In this chapter, I summarize key considerations in restoration project planning and refer readers to Rieger, Stanley, and Traynor (2014), who provide extensive detail on the planning process. I discuss the *adaptive management* cycle and *monitoring* in detail in chapter 4 and discuss methods for restoring specific *abiotic* and *biotic* conditions in subsequent chapters.

Stakeholder Engagement

Stakeholders are individuals, groups, or organizations that have a vested interest in a restoration activity, usually because they have something to gain or lose from it. Typical stakeholders include *natural resource managers*, industry groups, neighboring landowners, farmers, indigenous groups, government agencies with jurisdiction over the land or waterway, recreational users, and scientists (see the Sacramento River case study for an example list of stakeholders). Most projects have a diversity of stakeholders, and larger projects typically have more stakeholders than smaller ones. Some mandated projects, such as mine reclamation, may have only a few stakeholders, such as the government and the company addressing the cleanup.

Although it can be difficult to address all stakeholder desires and concerns, establishing an open, respectful, and transparent forum for stakeholder participation makes it more likely that multiple ecological and social needs can be met, thus increasing the likelihood of long-term restoration success. Furthermore, engaging stakeholders in restoration efforts is important for gaining support from public funding and for collaborative learning.

Stakeholders need to be engaged early and often throughout the planning process. Identifying conflicts and agreeing on goals up front allows project planners to avoid pitfalls that can plague otherwise well-planned projects. Project leaders should make use of stakeholder knowledge, expertise, and experience and should clarify potential misunderstandings as soon as they arise (Walker, Senecah, and Daniels 2006). Leaders who encourage a diverse group of stakeholders to participate in early decision-making and throughout the project usually find that their projects not only run more smoothly, but that the outcome is more successful. For example, Derak et al. (2018) describe a successful woodland restoration project in northern Morocco that included more than sixty stakeholders from a diversity of ages, genders, education levels, and socioprofessional backgrounds in the planning, implementing, and monitoring process. There, stakeholders started by openly discussing land use options and prioritizing restoration actions that maximized *ecosystem services* and then were involved in planning the logistics and planting of the project. Two years after project implementation, tree survival and natural vegetation establishment were high, there had been no site vandalism, and project participants reported on the positive benefits of social involvement and learning new information about forest *management* and restoration.

In contrast, early efforts to restore riparian forest along the Sacramento River in California were met with opposition from farmers who were concerned about increased flooding and the spread of plants and animals considered to be agricultural pests, as well as loss of farmland (Sacramento River case study). This challenge led to the local advisory board voting to reduce the area designated for riparian forest conservation and restoration by half and to local governments instituting stricter regulations to protect landowners from potential negative effects from nearby restoration sites. In recent years, conservation organizations working to restore the Sacramento River have set up stakeholder advisory committees to facilitate dialogue throughout the restoration planning and implementation process, which has both advanced the science and improved restoration success. The Asian Mangrove case study provides a clear example of how meaningful and ongoing engagement with the local community is a primary factor determining restoration outcomes. When the local community was involved throughout the planning and implementation process, survival of planted trees was generally high, but in cases in which the trees were planted by outside organizations without engaging the community, ongoing maintenance was minimal and tree survival was low.

Stakeholders can be involved in a variety of ways. It often begins with public meetings or the establishment of an ongoing forum for the exchange of information with clear representation from all specific stakeholder groups. In some cases, a more formal, structured decision-making approach that involves a range of analytic tools to guide decision makers through a transparent planning process is used. For example, Guerrero et al. (2017) describe a multistep planning process to set priorities for public funding to restore vegetation across eight hundred conservation parks in Queensland, Australia. The first step was to survey a large number of stakeholders to understand their priorities, which was followed by a workshop with a subset of key stakeholders to synthesize and refine these priorities. As a final step, they used a decision-support tool to compare trade-offs in goals and prioritize investments in restoration actions.

Garcia (2017) describes another approach using simulation games to work toward consensus between stakeholders to sustainably manage forests in the Congo Basin. Stakeholders (e.g., Forest Stewardship Council certified logging companies, international nongovernmental organizations, local community representatives, government officials) played games that asked them to make decisions about the types of forest management actions to apply and which areas to protect; then they compared outcomes for timber profits, local communities, and forest conservation. Exploring various scenarios helped illustrate concepts and outcomes and resulted in several points of consensus by the end of the three-day workshop.

Regardless of the approach, Walker, Senacah, and Daniels (2006) emphasize the importance of making the process open to as many stakeholders as possible, valuing the input of each participant, and incorporating stakeholders' ideas meaningfully and transparently. Encouraging groups and individuals with diverse interests to talk face to face about decision-making takes time, but it can ensure that a project is set up to succeed.

Goal Setting

A critical step early in the planning process is to state the *goals* of the restoration project as specifically as possible (fig. 3.1). This approach may sound obvious, but many projects skip this important step, moving forward with unclear or vague project goals that provide little guidance to those responsible for their implementation and postproject evaluation (Lockwood and Pimm 1999). Goals that are overly broad can lead to conflicting expectations among stakeholders. For example, a stated goal to "restore native grassland" could be interpreted in several different ways, such as restoring grassland to (1) increase the cover of native plant species, (2) enhance the population of an endangered native grassland insect, (3) reduce plant biomass to minimize fire risk, (4) increase carbon content of soils, (5) enhance the cover of native grassland plants that are food sources for indigenous peoples, (6) some combination of options 1 through 5, or (7) a different goal entirely.

Restoration projects can have multiple goals for restoring a given *habitat* type, but it is rarely possible to simultaneously maximize them all because they often conflict. For example, a grassland project to increase plant growth to maximize carbon storage could also increase fire risk. Likewise, the plant species desired by an indigenous group may not be the same species needed by the endangered insect. Hence, it is important to establish detailed goals early in the planning process to minimize conflicts, select appropriate restoration strategies, and be able to evaluate success at a later point.

Many projects have social goals that should be stated clearly; they often include job creation, environmental education, increasing recreational opportunities, and community involvement. The Greenbelt Movement in Kenya led by Nobel Prize–winner Wangari Maathai encouraged women to grow seedlings and plant trees with the goals of minimizing erosion, improving water storage, and providing food and firewood, as well as empowering women and promoting participatory democracy (Greenbelt Movement n.d.). Similarly, many urban restoration projects aim to reconnect and educate both children and adults about their local natural history.

Once general goals that provide direction and vision for the project have been agreed upon, they must be paired with measureable *objectives* (Gann et al. 2019), commonly referred to as SMART goals or performance criteria.

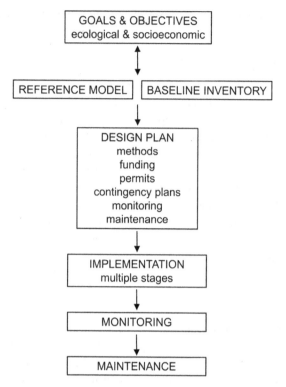

Figure 3.1. Steps in restoration project planning and implementation.

Objectives must be specific (S), be measurable (M), be achievable (A) given existing constraints and available resources, be relevant (R) to evaluating whether the broader goals are achieved, and have a clear time frame (T) for completion. Restoring native plant cover in grasslands is an acceptable goal, but it does not provide any information on how much cover and by when the native cover should be restored, both of which are important to determine whether the project was successful. An objective paired with this goal might be to achieve 30 percent native plant cover within three years and 50 percent native plant cover within ten years. Similarly, a social objective might be a minimum of ten field visits by local middle and elementary schools each year. Table 3.1 provides additional examples of goals and associated objectives for the Younger Lagoon case study.

Goals and objectives should be set early in the planning process. To ensure that they are achievable and reasonable, this step should be done concurrently with developing a reference model and conducting a baseline inventory and constraints to restoration at the project site. As the planning

Table 3.1. Selected Goals and Objectives for Restoration of Coastal Prairie and Freshwater Wetland Habitat at Younger Lagoon Reserve[1] on the Central California Coast

Goal	Objective	Evaluation time	Action if target not met
Restore native coastal prairie vegetation	4 or more native plant species established per transect and ≥ 10% native cover	2 years post-planting	Perform supplemental planting and weed control
	6 or more native plant species established per transect, ≥ 25% native cover, evidence of natural *recruitment* of native species	5 years post-planting and thereafter	Perform supplemental planting and weed control; consult scientific advisory committee
Remove all high-priority weeds	No high-priority weeds on site	3 years from start of plan	Continue weed control and consider alternative methods
Restore wetland hydrology	Restore hydrological flow from wetland 2 to wetland 1	1, 2, and 3 years following installation of diversion	Enhance diversion structure
Provide interpretation opportunity for visitors to nearby marine discovery center	Offer docent-led tours twice monthly	Ongoing	Increase frequency of tours
Protect native habitat, wildlife, and research and restoration efforts	Eliminate all nonservice domestic animals from the site	Ongoing	Increase outreach and education to reserve visitors; coordinate with police department on violations

[1]See Younger Lagoon case study for more details.

group agrees on reference models and conducts initial research and site evaluations, new information or unforeseen constraints may require revising the original project objectives.

Develop a Reference Model

Setting goals and specific objectives requires choosing a *reference model* or reference *ecosystem* for the habitat being restored. A reference model represents the approximate target for restoration and should be based on an understanding of the biophysical processes and ecological interactions in a given ecosystem (chaps. 5 and 6; Clewell and Aronson 2013). McDonald et al. (2016, 6) clarify that the reference model "aims to characterize the condition of the ecosystem as it would be had it not been degraded, adjusted as necessary to accommodate changed or predicted biotic or environmental conditions"; this model is developed from multiple sources of information about past, present, and anticipated future conditions at the site and similar sites in the region. Project managers must ask themselves whether the ideal reference model approximates how the ecosystem was a century ago, the best available nearby *reference sites*, or a mix of species that are well adapted to current and future conditions. For some goals, such as improving water quality, there are published standards of what concentrations of certain chemicals are safe for humans and some fauna species, which can inform objectives. In most cases, however, as discussed in chapter 2, selecting an appropriate reference model to inform project objectives is complicated and involves subjectivity. Drawing on many different sources of information and soliciting input from knowledgeable stakeholders, *practitioners*, scientists, and others with expertise in the ecosystem type being restored can help in this process.

One potential source of information for the reference model is historical data. If the site was disturbed recently, then data from the site prior to that *disturbance* are particularly useful. In many cases, however, such data are not available. Determining what an ecosystem looked like fifty years ago, one hundred years ago, or longer requires ecological detective work to piece together information from varied sources (Egan and Howell 2001; Clewell and Aronson 2013). Written documents, such as field survey notes, species lists, natural history journals, research papers, floras and faunas, and unpublished reports, can all be valuable sources of information.

Oral histories that document *traditional ecological knowledge* and *local ecological knowledge* (also referred to together as "indigenous and local knowledge") of people who lived in a region for long periods of time can be informative (Uprety et al. 2012; Potts et al. 2018; Gann et al. 2019). These people often have detailed knowledge of the historic species composition and how to manage the ecosystems to facilitate *recovery* based on their own experience. For example, Wehi (2009) reviewed more than twenty-five hundred ancestral sayings of the Maori people in New Zealand and found that 9.4 percent of them referred to terrestrial plant species, thus providing valuable information on their habitat requirements, interactions with animals, and past management practices. Mamun (2010) describes how local fishers' knowledge of fish movement and habitat selection can be used to improve coastal habitat restoration in Bangladesh. Drawing on local ecological knowledge provides important insights into the ecosystem and also engages local communities in the restoration process. Historical photographs, maps, and artwork provide a visual representation of the site at an earlier time. Museum specimens and pollen samples offer evidence of what species were present at a site in the past, and tree rings provide insight into previous climatic conditions. Some people even use packrat middens, which can preserve seeds and animal bones for hundreds or even thousands of years in arid areas. None of these sources provides as complete a picture of an ecosystem as contemporary data, but when available, they can all help inform the reference model.

Most restoration projects use data from relatively intact, nearby *reference sites* with conditions similar to those where the restoration project will be implemented as a source of information for the reference model; these sites are typically selected by consulting with experts or looking at historical aerial photographs to select sites that have received minimal human impact. Using data from multiple reference sites, ideally repeatedly collected over multiple years, to incorporate both natural spatial and temporal variation when defining restoration objectives is recommended (White and Walker 1997). Reference sites often provide the most realistic target that is achievable given irreversible changes that have occurred, such as species invasions and *extinctions*. For example, even highly diverse California grasslands usually have greater than 50 percent cover of nonnative grasses and forbs, meaning that an objective of 100 percent *native species* cover is not realistically achievable in most cases.

Given global climate change and other anthropogenic impacts, it is becoming increasingly important to consider possible future conditions when developing a reference model. Future states are difficult to anticipate, so to ensure that the project will succeed under future conditions, models of predicted future climatic patterns, hydrologic flows, and species distributions are often used to inform the reference model and project design plan. As one example, Veloz et al. (2013) use models of sea-level rise to guide future wetland restoration in the San Francisco Bay and most effectively restore habitat for several tidal marsh bird *species of concern*. Restoration practitioners are increasingly considering future climatic conditions in selecting species and *ecotypes* for restoration (chap. 9). Likewise, different strategies are being considered when planning for enhanced survival in coral restoration projects, such as selecting ecotypes of coral or their algal associates that are adapted to higher temperatures and acclimatizing them to elevated temperature prior to outplanting (van Oppen et al. 2015).

Analysis of Existing Conditions

Another important early step in project planning is to analyze existing conditions that will affect and constrain restoration efforts both within and outside the site and resolve these constraints to the degree possible; this process should be done in parallel with goal setting and defining a reference model. A first step is to conduct a *baseline inventory* to assess the current abiotic and biotic conditions of the site to be restored (see fig. 3.1). This information serves as a reference point for restoration planning and to evaluate progress later in the restoration project. Abiotic conditions that are typically assessed include soil and water chemistry, hydrologic flows, topographic variability, and microclimatic conditions. Biotic conditions commonly include the composition and abundance of both native and *nonnative species* on the site, often with an emphasis on *focal species* for the restoration project. Detailed maps of the site that incorporate spatial heterogeneity in abiotic and biotic conditions (e.g., soil type, depth to groundwater, species distributions) will help guide the design of the project. For example, riparian restoration plans along the Sacramento River include baseline data on within-site variability in soil type and depth to groundwater; grassland and savanna species are used in areas with sandy soils and deeper groundwater, whereas riparian forest species are planted in fertile soils with shallow groundwater (Sacramento River case study). It is also important to assess factors outside the restoration site that influence recovery, such as inputs of water and nutrients, and potential sources of faunal and floral *propagules* in the landscape (chaps. 5 and 6).

The baseline inventory of conditions within and surrounding the site will help inform what goals and objectives are feasible and the degree of intervention needed to achieve them. For example, if abiotic conditions are only moderately disturbed and the desired species are present on-site or in nearby ecosystems, it may be most cost-effective to use a *natural regeneration* approach to restore the system. On the other hand, if *invasive species* are widespread, planners will need to select methods to remove those species and reintroduce native species.

Assessing ongoing and potential future constraints at the outset can guide decisions about how to address, compensate, or adapt to the specific limitations of a site (Gann et al. 2019). It is critical to resolve major existing stresses to the ecosystem (e.g., elevated nutrient inputs from outside the site, altered *hydrology*, nearby sources of invasive species) prior to undertaking restoration to increase the likelihood of restoration success and reduce the need for ongoing *ecosystem maintenance*. When it is not feasible to resolve major constraints, full recovery is unlikely, and the project goals should reflect that point. For example, restoring the historical *hydrological regime* and river channel pattern in urban streams is rarely feasible given ubiquitous human water withdrawals, development adjacent to rivers, and concerns about flooding. Therefore, such projects usually focus on improving the habitat quality immediately adjacent to the river, water quality, and green spaces for urban populations rather than aiming to restore historical flows (Riley 2016). Some minor constraints on the system can be addressed with restoration methods. For example, mammal or bird herbivory on recently planted seedlings can be reduced with various types of plant protection devices (chap. 9). In addition, clear signage, education programs, and community involvement can reduce human vandalism of restoration projects.

Design Plan Considerations

Once the reference model has been developed and the existing conditions have been evaluated, the goals and objectives need to be finalized, all of which will guide the next step: the design plan (see fig. 3.1; Rieger, Stanley, and Traynor 2014). The project design plan details the guidelines, methods, technical drawings, and timeline for implementing the project, as well as who is responsible for taking the various actions described in the plan. The initial draft serves to clearly communicate the plan to others and to solicit feedback for improvement. After incorporating feedback, the final draft guides the restoration project. Rieger, Stanley, and Traynor (2014) provide a detailed explanation of the process and helpful resources for writing project design plans (see book website for design plan examples).

Selecting Restoration Methods

Many restoration techniques could be used in any given project, so it is important to evaluate the potential success of alternative methods in achieving the stated project goals and objectives. Doing so means reviewing the relevant literature, talking to individuals with expertise in various fields, and soliciting the experiences and feedback of stakeholders. In systems in which there has been extensive past restoration, the *best management practices* (i.e., the most effective and feasible restoration practices) may be well established, but more often there is a range of options that will need to be evaluated and tested to determine which is the most effective for the site conditions and goals of the project. For example, whether to use manual removal, herbicides, controlled burns, or some other method to remove *invasive* plant species (chap. 8) depends on the ecosystem type, what actions are permissible at the site, the amount of labor available, and a host of other factors. The design plan should provide a brief rationale for and detailed description of the restoration activities that will be undertaken to improve the abiotic and biotic conditions on the site. The plan should also describe any potential unintended effects of the selected restoration actions and describe steps that will be taken to minimize potential negative effects (Gann et al. 2019). Finally, the plan should clearly outline guidelines for the sourcing and genetics of plant or animal species that will be actively introduced (chaps. 9 and 10).

Contingency Plans, Monitoring, and Maintenance

Given most budget constraints, restoration projects frequently focus on the first one to three years. Project leaders commonly fail to consider what happens if things do not go as planned and how the project will be maintained over the longer term. But restoration projects rarely proceed exactly as planned for a host of reasons, such as natural disturbances (e.g., drought or flooding), changes in funding support or political will, unforeseen site conditions, and failure of restoration methods (Rieger, Stanley, and Traynor 2014). Moreover, ecosystem recovery takes decades to centuries, so rarely will restoration projects be successful in the long term without ongoing *monitoring* and maintenance.

Restoration practitioners should therefore take an adaptive management approach, which refers to improving ecosystem management by learning from implementing projects, evaluating their outcomes, and taking corrective actions (chap. 4; Walters 1986), and plan for contingencies when things inevitably go awry. Thus, the design plan should outline a monitoring strategy and discuss how monitoring data will be used to guide corrective actions (chap. 4). Table 3.1 illustrates a few examples; for example, if certain native cover and diversity objectives are not reached by a certain time, then additional planting or seeding needs to be done. For projects that involve seeding or planting, establishment rates are highly dependent on amount and timing of precipitation, so to increase the likelihood of successful vegetation establishment, it is wise to plant over multiple years (Wilson 2015). Obviously, it is not possible to consider all possible outcomes, but it is helpful to consider the most likely risks and how they would be addressed.

The design plan needs to take a longer-term view and should discuss how the project will be maintained beyond the initial implementation stage. Some questions that should be addressed in the design plan include:

- What is the long-term land tenure situation? Is it secure?
- Who is responsible for adaptively managing and maintaining the project shortly after implementation, as well as over the long term?
- When will project objectives be evaluated and potential corrective actions be taken?
- What is the funding source for ongoing project management?

These questions are frequently neglected, leading to many restoration projects having only temporary effects on ecosystem recovery. Many mangrove restoration projects in Asia illustrate this common restoration problem, where millions of dollars have been spent on restoration activities such as tree planting without careful prior planning or consideration of how the trees, and the ecosystem as a whole, will be managed over the long term (Asian Mangrove case study). Often, such projects have a high failure rate, wasting the initial investments.

Permits

Another critical part of the restoration planning process is to identify any legal constraints governing the restoration site and to acquire the necessary permits. For projects that occur in countries that have strict environmental regulations (chap. 11), obtaining permits can be time-consuming, so much so that some large consulting firms have employees whose sole job is to secure permits. Permits from local, state, and federal agencies may be needed for projects that impact air and water quality, water flow, sensitive habitats, or *species of concern*; disrupt traffic; use herbicides or pesticides; or require seed collection from public lands. Therefore, it is essential to begin the process of obtaining permits early in the planning phase. Table 3.2 provides an example of the list of permits needed for a relatively small coastal habitat restoration project in California. The type and extent of permits required depend on the ecosystem type and regulatory framework of the project location, so knowledge of local and federal regulations and the associated permitting process for the project site is essential.

Resources and Budget

Selecting project-appropriate methods requires evaluating not only project goals and constraints, but also the resources available. A major obstacle to many restoration projects is the high cost (chap. 12), so the design plan must include a detailed budget and clearly indicate who will pay for the project before any implementation begins (see fig. 3.1). In addition to financial

Table 3.2. Permits Required for Restoration[1] of Coastal Prairie and Freshwater Wetland Habitat at Younger Lagoon Reserve[2] on the Central California Coast

Permitting agency	Law requiring permit	Reason for permit
California Coastal Commission	California Coastal Act	Building construction and restoration work in the coastal zone
California Department of Pesticide Regulation	Title 3 of the California Code of Regulations	Use of herbicides for restoration or scientific purposes
California Department of Fish and Wildlife	Fish and Game Code Section 1002 and Title 14 Sections 650 and 670.7	Scientific collection and handling of protected plant or animal species
California State Parks	California Code of Regulations	Seed collection from reference sites managed by California State Parks
California State Water Resources Control Board	Section 1251 of the US Clean Water Act	Stormwater discharge during construction and restoration projects
US Army Corps of Engineers	Section 404 of the US Clean Water Act	Presence of wetlands
US Fish and Wildlife Service	Endangered Species Act	Potential habitat for the California red-legged frog[3]

[1]Additional permits are required for scientific research on restoration.

[2]See Younger Lagoon case study for more details.

[3]California red-legged frog (*Rana draytonii*) is a threatened species.

resources, planners must consider what other resources are needed and who is responsible for providing them. Where will the supply of plants and seed come from? Is there access to irrigation water on-site? If volunteers are involved, who will recruit and coordinate them? Because unexpected costs are a reality of restoration projects, it is important both to have additional contingency funding built into the plan to provide some flexibility and to identify who will be responsible if costs exceed the budget.

Timeline

Design plans must give careful thought to the timeline for project implementation, monitoring, and maintenance. Most restoration projects take a great deal of advance preparation to ensure that resources are available at the time they are needed. For example, projects involving revegetation may require multiple years to collect sufficient native seed, grow the seed in the greenhouse, and prepare the plants for outplanting before revegetation can

Table 3.3. Timeline for a Small-Scale Grassland Restoration Project[1] at Younger Lagoon Reserve[2]

Activity	At least 1 year prior to project	Aug.– early Oct.	Oct.– Nov.	Dec.	1 week prior to planting	Mid-Dec.– Feb.	March	Ongoing
Get permits	★							
Begin weed control	★							
Collect seed	★[3]							
Germinate seeds in flats		★						
Transplant seedlings to pots			★	★				
Install rabbit fencing around site				★				
Arrange for labor and assemble tools for project				★				
Final weed control					★			
Harden plants					★			
Spread mulch					★			
Plant seedlings					★	★		
Replace dead plants							★	
Spring and fall mowing to control weeds								★

[1]Timelines for larger projects, particularly for those that include earth-moving, are much more complex.

[2]See Younger Lagoon case study for more details.

[3]Because seed set is highly variable from year to year and some seed needs to ripen before germination, it is best to start seed collection in the spring and summer the year before germinating seeds and continue collecting in the spring and summer immediately prior.

begin (table 3.3). Working through the planning process with stakeholders and securing permits and funding may take even longer. Therefore, it is important to thoroughly review the order and timing of each step so that all the necessary permissions and resources are available when needed to implement the project. Inevitably, there are project delays due to unexpected circumstances, which will require some flexibility in the schedule.

Implementation

Once the design plan is completed, the next phase is implementation (see fig. 3.1). When possible, it is best to use a "staged restoration" approach

(Bakker et al. 2018), first testing methods, particularly those that are novel, at a small scale before progressively scaling up. For revegetation projects, pilot efforts often consist of small-scale plantings of many species to select the subset that performs best under local site conditions. A good example of a large-scale staged restoration approach is the Kissimmee River restoration project in Florida, where the 90-kilometer-long channel was restored to a 166-kilometer-long meandering river (Kissimmee River case study). The first steps were to build small-scale physical models for the restoration (Koebel and Bousquin 2014) and to use numerical models to simulate and compare the hydrologic conditions of three restoration options. Then a small section of the river was restored and results closely monitored before the full project was completed in four stages. However, the staged restoration approach can be challenging to implement if restoration funding is only available for one to a few years. Regardless, conducting pilot studies will likely pay off in the long run because they often save considerable money by identifying the most cost-effective methods and enhancing the likelihood of success over the long term.

Project implementation requires careful coordination to ensure the proper timing of labor, equipment, and materials at the site (Rieger, Stanley, and Traynor 2014). Hiring staff or recruiting volunteers before permits have been acquired or the plants are ready to outplant can lead to frustration and delays. Prior to implementation, it should be clear which staff members are responsible for which tasks, and all staff and volunteers should be well trained on correct methodologies. Supervisory staff should visit the project site frequently to ensure that methods are being implemented correctly and to assess when the project plan needs to be adjusted. Often, some modifications to the restoration methods will be necessary, and these changes should be made in consultation with the project designer and any groups with regulatory oversight. It is important to keep detailed records of actions taken and their associated costs, particularly when there are changes to the design plan, and to share progress and changes to plans with stakeholders frequently. This information, in combination with monitoring data, can help document how successful different approaches are and thereby inform future restoration efforts (chap. 4).

Recommended Reading

Clewell, A. F., and J. Aronson. 2013. *Ecological Restoration: Principles, Values, and Structure of an Emerging Profession*. Washington, DC: Island Press.
>   Provides a broad overview of restoration ecology, including topics such as why to restore ecosystems, how to define the reference system, and how to plan for ecosystem restoration projects.

Egan, David, and Evelyn A. Howell. 2001. *The Historical Ecology Handbook: A Restoration-ist's Guide to Reference Ecosystems*. Washington, DC: Island Press.

Provides extensive information on collecting and using different types of historical data to characterize the reference model.

Rieger, John, John Stanley, and Ray Traynor. 2014. *Project Planning and Management for Ecological Restoration*. Washington, DC: Island Press.

Provides a detailed discussion of the nuts and bolts of restoration project planning and implementation from a practitioner's perspective.

4

Monitoring and Adaptive Management

Billions of dollars and millions of hours of paid and volunteer labor are spent to restore damaged ecosystems every year. But many of these projects are short-lived, and, when they last, we rarely know whether the original objectives were achieved. Why? In part, it is because most restoration projects lack a robust monitoring protocol and a plan for adaptive management, both of which are critical to determine whether restoration efforts are succeeding and, if not, to trigger corrective actions. Even in cases in which design plans call for monitoring and adaptive management, most monitoring plans fall short of being able to evaluate success because the plan did not specify measurable objectives or a timeline for their measurement (chap. 3), the variables monitored do not correspond to the objectives, or there is no clear plan for what to do with the data once they are collected (Elzinga, Salzer, and Willoughby 1998; Lindenmayer and Likens 2018). Bernhardt et al. (2005) reviewed more than thirty-seven thousand river restoration projects in the United States, and only 10 percent of them had evidence that monitoring had occurred; for many of them, the monitoring programs were not well designed to evaluate the objectives of restoration actions.

In this chapter, I provide an overview of adaptive management and discuss general considerations and recommendations for developing cost-effective restoration monitoring protocols. I do not discuss monitoring parameters for individual ecosystem types or specific project objectives in detail; rather, I refer readers to additional resources that thoroughly describe these.

Monitoring is more than collecting data. It is the systematic and orderly gathering of data over a period of time to evaluate whether specific project objectives are achieved (Holl and Cairns 2002). Monitoring helps identify problems as they arise and save money in the long term because taking corrective actions is less costly than discovering and solving problems long after they occur (Chaves et al. 2015). In addition, monitoring one project often generates information that can improve the overall success and cost-effectiveness of future restoration efforts if the lessons learned are shared (Kondolf 1995). For example, between 1987 and 1991, 569 swift foxes (*Vulpes velox*) were reintroduced to areas where they had been *extirpated* in central Canada, and their survival and movement patterns were monitored using radio collars. The results showed that foxes released in the fall had more than twice the survival rate of those released in spring, which was the opposite of what managers had predicted; using this information has since increased the success of subsequent *reintroduction* efforts (Carbyn, Armbruster, and Mamo 1994).

Monitoring helps determine whether specific objectives have been achieved, but it will not necessarily explain the underlying causes of the success or failure of restoration strategies. Controlled experiments are the best way to determine the reasons for a given restoration outcome and to rigorously compare the efficacy of different restoration strategies. For example, if restoration plantings grow slowly, then it is difficult to identify the cause (e.g., inappropriate species selection, lack of nutrients, water stress, or *competition*) without experiments designed to control causal variables. Hence, I encourage resource managers to collaborate with scientists to design and implement restoration experimentally. Doing so can take a range of forms, from replicated experiments testing individual *management* actions (e.g., invasive nonnative control and seeding methods as in the Younger Lagoon case study) to resource managers implementing two or three broad restoration approaches in a site followed by systematic monitoring. Another valuable approach is to synthesize monitoring results from multiple restoration projects that use similar methods. Alexander and D'Antonio (2003) compared twenty sites where land managers had used a range of methods to control two invasive shrubs in California, French broom (*Genista monspessulana*) and Scotch broom (*Cytisus scoparius*). They found that repeated hand pulling and burning were the most effective methods and that the degree to which native vegetation recovered varied along a coastal to inland rainfall gradient.

Adaptive Management Cycle

Adaptive management is an approach to restoration and land management that focuses on learning by doing; monitoring data are used to inform corrective actions on the current project and increase the likelihood of success for future restoration efforts. Adaptive management sounds simple, but it requires a series of carefully planned steps (fig. 4.1).

Define Clear Goals and Specific Time-Limited Objectives

To be able to evaluate restoration success, it is critical to establish clear *objectives* (also called performance criteria) at the outset that include a measurable variable, a desired direction of change (e.g., increase, decrease, maintain), and a time frame (chap. 3; Elzinga, Salzer, and Willoughby 1998). A statement such as "increase native plant cover" is not a useful objective because it does not provide information on how much the cover should increase or the time frame within which it needs to happen. A measurable objective might be "increase native plant cover to 70 percent within five years." Well-developed *reference models* (chap. 3) help set reasonable objectives.

Select Monitoring Parameters and Corrective Action Trigger Points

The next step is to select specific *parameters* (variables that correspond with each objective) and the method and timeline for measuring those parameters. During planning, it is important to determine how much variation from the stated objectives is acceptable or, in other words, the degree of similarity to a reference system required for the project to be considered a success. If objectives are not met, then specific corrective actions should be triggered (see fig. 4.1). For example, if a prespecified level of native plant cover is not achieved, then further planting is required (see table 3.1).

Conduct Baseline Inventory

During the planning process, a *baseline inventory* is needed to assess the current abiotic and biotic conditions at the site (chap. 3). These data inform the restoration planning process and serve as a comparison to evaluate whether restoration leads to significant improvement of site conditions over time.

Monitor, Analyze Data, and Determine Whether Corrective Action Is Needed

Data collection should begin soon after the start of restoration and be repeated at the intervals specified in the monitoring plan. For example, Chaves et al. (2015) outline a schedule of monitoring progress on Atlantic

Figure 4.1. The multiple steps in the adaptive management cycle. Figure by A. Calle.

forest restoration at three, five, ten, fifteen, and twenty years postrestoration (Atlantic Forest case study). Collecting and evaluating monitoring data in a timely manner are essential to determine whether the restoration site is moving along the desired *trajectory* and the objectives are being reached, and if not, to generate early warnings that corrective actions are needed. Yet it is not uncommon for collected data to accumulate over time without being analyzed due to lack of time or statistical expertise.

It is important to have resources available to undertake corrective actions, a key component of adaptive management. Otherwise, becoming aware of restoration shortcomings will not translate into actions to move the project back on course. Therefore, corrective actions should be agreed upon and budgeted for ahead of project implementation so that if a trigger point is reached, corrective actions can be taken. If the project meets the objectives at a given time, then managers should continue monitoring and following the adaptive management cycle (see fig. 4.1).

Although many restoration projects propose to use an adaptive management approach, only a small fraction of them actually do. Why? First and foremost, funding for restoration projects is limited and often constrained to one to a few years (chap. 12), which means that *natural resource managers* often do not have the funding for monitoring and taking corrective actions over the time scale that is needed. Moreover, they may not have the technical expertise to measure certain parameters or the statistical training to analyze the data. In some cases, the actions needed to achieve the desired outcomes (e.g., reducing river water withdrawal upstream) are not within their control. Finally, monitoring often shows that projects are not going as planned, and nobody wants to admit that their project was unsuccessful. Despite these real obstacles, following the adaptive management cycle helps reduce uncertainties in the restoration methods and is critical to improving restoration success.

Selecting Monitoring Parameters and Methods

The most important criterion in selecting monitoring parameters (also referred to as variables) is that the parameters evaluate whether the objectives have been met. For example, if one of the project objectives is to decrease phosphorus (P) concentrations in a lake by 20 percent within three years, then P concentrations must be measured over time. If the project aims to create a certain amount of jobs, then the total number of people employed should be tracked. This approach may seem obvious, but in a surprising number of cases, the variables monitored do not match the objectives or the monitoring focuses on whether certain restoration actions were undertaken (e.g., a certain number of trees were planted), rather than whether the desired ecological or socioeconomic goals and objectives were achieved (May, Hobbs, and Valentine 2017). For example, Murcia et al. (2016) reviewed more than one hundred forest restoration projects in Colombia and found that a primary goal of 89 percent of the projects was restoring potable water supply. In sharp contrast, 96 percent of the projects measured short-term vegetation variables, such as survival and growth of planted trees, vegetation ground cover, and erosion control. This difference is likely because the measured vegetation parameters are much easier to monitor than other parameters. Nonetheless, measurements of water quantity and quality are needed to determine whether the water supply goal is being achieved.

Measured parameters can be physical attributes (e.g., channel width, soil compaction) or biological factors (e.g., abundance, richness, or composition of floral or faunal groups of interest, maintaining a specified population size of a *focal species*; table 4.1). It is also important to monitor social or socioeconomic goals of restoration (Martin and Lyons 2018); parameters might include the number of recreational users per year, neighbors' attitudes toward the project, community participation in planning meetings or restoration work days, or number of homes that experience reduced flooding risk.

There are many detailed monitoring manuals for specific organism or ecosystem types, such as measuring plants and animals (Elzinga, Salzer,

Table 4.1. Commonly Measured Parameters for Evaluating Restoration

Physical habitat and functions
- *Topography*—slope, elevation, gullies, water depth
- *Hydrology*—quantity and rate of surface or groundwater flow throughout the year, timing of discharge
- River channel form and in-channel structure—*sinuosity*, width-depth ratio, pools and *riffles*, streambed substrate
- Soil movement/erosion—sediment flux
- Water quality—turbidity, dissolved oxygen, nutrient and toxicant concentrations, pH
- Temperature—air, soil, water
- Light—shading of terrestrial or riparian habitat, penetration in water
- Soil physical characteristics—*compaction*, water-holding capacity, infiltration, texture
- Soil chemical characteristics—pH, *organic matter*, nutrient and toxicant concentration
- Nutrient cycling—air, water, soil, e.g., nitrogen fixation and mineralization
- Disturbance frequency/intensity—e.g., fire, flood
- Connectivity with adjacent habitats

Biota
- Vegetation and faunal *community composition*—abundance or cover of *native species*, certain focal or rare species or certain functional groups, species richness and evenness, presence or abundance of invasive *nonnative species*
- Vegetation structure—cover of different vegetation layers, height, stem density, basal area of trees, biomass
- Faunal health—toxicant level, malformations
- Reproduction and mortality rate—flora, fauna
- Movement of fauna in and out of restored habitat and home range
- Undesirable species—presence or abundance of predators, diseases, *invasive species*

Socioeconomic
- Recreational opportunities
- Job and *livelihood* opportunities
- Hazard risk reduction
- Income generation—e.g., harvest of timber, fuelwood, and other products, recreation, fisheries
- Educational programs
- Participation in volunteer programs
- Participation of private landowners or other stakeholders
- Aesthetic values
- Compliance with government plans and policies

Sources: Westman 1991; Elzinga, Salzer, and Willoughby 1998; Thayer et al. 2005; Palmer, Hondula, and Koch 2014.

and Willoughby 1998; Morrison 2009), coastal habitats (Thayer et al. 2005), rivers (Palmer, Hondula, and Koch 2014), and grasslands, shrublands, and savannas (Herrick et al. 2005). Even when monitoring methods are well established, it is helpful to test them to determine whether a specific monitoring method is feasible for a given site, to assess the level of monitoring expertise needed, and to work out logistical details. These steps can be done when conducting the baseline inventory.

Increasingly, remotely sensed data from satellites, airplanes, and unmanned-aerial vehicles (drones) are being used to monitor a range of variables, such as vegetation cover, vegetation structure for bird habitat, and lake water clarity (Zahawi et al. 2015; Abdullah et al. 2016; Dörnhöfer and Oppelt 2016). Remote-sensing tools have the potential to reduce the resources needed to monitor large areas and difficult-to-access locations, but require technical tools and skill in analyzing spatial data.

When developing a monitoring plan, it is tempting to propose to monitor many parameters. Given the reality of limited monitoring budgets, it is best to carefully select the most relevant and cost-effective measurement parameters that evaluate whether the objectives have been achieved. Regardless of the system or project, there are a few important factors to consider while developing monitoring plans (Holl and Cairns 2002).

First, simpler protocols are better so that the data can be collected by people without highly specialized training. The feasibility of measurements, including time, cost, and practicality, must be weighed against their importance for judging the success of achieving specific project objectives. Some measurements, such as monitoring toxic substances in water or tracking faunal movement and use of habitat, may require specialized equipment, be expensive, and require extensive training. If the measurements are important to evaluate specific restoration objectives, then they should be taken, but simpler methods are preferable.

Second, methods should be repeatable without high variability in measurements by different users (Elzinga, Salzer, and Willoughby 1998). For instance, estimates of plant cover often vary greatly from person to person, whereas using a point-intercept method (how many times a dropped pin intercepts different plant species) results in more comparable measurements among people. Third, when at all possible, standard techniques that are published in the literature should be used, thus allowing for comparisons between different projects.

Finally, caution should be taken in using indicator species (Holl and Cairns 2002; Lindenmayer and Likens 2018), which are mostly animal species or groups of species (e.g., birds, butterflies, amphibians) that are measured to represent specific physical, chemical, or biotic habitat conditions (e.g., water quality in aquatic systems, vegetation structure in terrestrial systems) or that are assumed to be representative of a wide range of species (e.g., using bird diversity to indicate insect diversity). Unfortunately, indicator species have proven to be problematic; none of the fifty-five groups that have been proposed as indicators consistently respond to specific environmental conditions or correlate with other groups across a range of sites or habitat types (Lindenmayer and Likens 2018). Moreover, plants and animals respond to many habitat and environmental changes, and determining the specific drivers of abundance changes is complicated and often poorly understood. Of course, it is essential to monitor a specific group or species if their recovery is one of the goals of the project. If indicator species are used, then there should be ample data showing a clear link between the group being monitored and the desired restoration goal they are being used to indicate, such as monitoring aquatic insects or fish that are sensitive to pollutants to indicate water quality (Herman and Nejadhashemi 2015).

Additional Considerations in Developing a Monitoring Plan

Below, I briefly discuss a number of additional considerations to develop a robust and informative restoration monitoring program. The recommended readings provide more detailed discussions of these topics.

Surveillance Monitoring

Most restoration monitoring focuses on compliance, or determining whether restoration projects are achieving the prespecified objectives at the desired time intervals. Another important type of monitoring is *surveillance monitoring*, which aims to catch unanticipated problems early on, before they have escalated to a level that is difficult to control. Surveillance monitoring is often used to periodically check the entire site for recently established *invasive species* of concern because the most effective way to control invasive species is to immediately remove small, newly established populations to prevent their further spread (chap. 8; Moody and Mack 1988). Likewise, doing a quick site check for recently established gullies or erosion hotspots and taking immediate actions to prevent further erosion can prevent greater expenses in the future. Surveillance monitoring differs from compliance monitoring in that it is less detailed but occurs at a larger scale to identify emerging problems.

Participatory Monitoring

Most often, monitoring is done by restoration project staff with specific expertise. In some cases, involving stakeholders in monitoring can have important benefits, such as promoting community participation and support, building trust, and reducing monitoring costs (Asian Mangrove case study; Evans, Guariguata, and Brancalion 2018). For example, Danielsen et al. (2011) compared forest biomass monitoring by community members and professional foresters in Tanzania and India. They found that appropriate training measurements taken by community members were similar to those taken by professionals, with the advantages of reducing costs and engaging community members in the project. Evans, Guariguata, and Brancalion (2018) review examples in several countries where community members successfully recorded monitoring data using smartphones. Participatory or "citizen science" monitoring has many benefits, but it requires skilled, paid staff to oversee volunteers and is only appropriate for collecting certain types of data that do not require a high level of technical expertise. Having paid staff is essential to recruiting and training volunteers on correct monitoring methods, ensuring quality control of the data, and compiling and analyzing data. Hence, participatory monitoring is not entirely free, and associated costs need to be included in the restoration project budget.

Timing and Frequency

The timing and frequency of compliance monitoring depends on the parameters being measured. In addition, the cost and labor availability will be a limiting factor; thus, it is important to allocate monitoring resources most efficiently to determine whether objectives are being met. For example, vegetation cover is often monitored once annually at the peak of the growing season, but sampling more than once may be necessary if focal species flower at different times. River flow is usually monitored during high and low flow events or at the time of year that is critical for a focal faunal species. Monitoring dissolved nutrient levels periodically throughout the year may be necessary where water bodies experience high temporal fluctuation, but is often concentrated after major rainfall events when nutrient runoff levels are highest.

Monitoring should be done more frequently shortly after project completion to determine whether restoration efforts are proceeding along the desired trajectory and thereafter at longer intervals. Typically, survival and growth of planted vegetation are measured annually for the first few years and then every two or three years thereafter. If the parameters measured fall below or above a trigger point that requires action (see fig. 4.1), then more frequent monitoring should be resumed.

Duration

Both restoration and monitoring time frames are typically short (e.g., two to five years). They are commonly limited by budget constraints, and timing is guided by the need to demonstrate compliance with regulatory standards. For example, Bayraktarov et al. (2016) reported that of 235 marine coastal restoration projects, 47 percent of projects were monitored for less than one year, 21 percent were monitored for one to two years, 21 percent were monitored for more than two years, and 11 percent had no information on duration. Clearly, monitoring for only a few years is insufficient to evaluate restoration success because most ecosystems take much longer to recover from disturbance (chap. 5). Moreover, restoration success can be short-lived or may reverse over time. A number of grassland restoration projects in California have appeared to be successful in the first two years following project completion, but were dominated by nonnative species by the third and fourth years (e.g., Holl et al. 2014).

Ideally, monitoring should continue until the ecosystem is self-regulating, meaning that the community composition, *ecosystem structure*, and *ecosystem processes* persist in the absence of management (e.g., irrigation or nutrient addition, removal of invasive species) that may have been necessary during the initial restoration efforts. Unfortunately, the time required for most ecosystems to achieve a self-regulating state may be well beyond what is financially and politically feasible for monitoring to continue. At a minimum, monitoring should occur until the final objectives are achieved and sustained for a few years thereafter. Data should also be collected over a sufficient time period to incorporate natural cycles of variation, such as precipitation.

Number and Spatial Distribution of Samples

Monitoring efforts aim to evaluate the success of an entire restoration project based on samples from a number of locations within a site. How many samples to take and how to distribute them throughout a site are common questions. Importantly, the sampling locations should be selected in an unbiased manner, either randomly or systematically (e.g., at specific distance intervals along a transect). There is a tendency to sample in areas that are easy to access or appear to be more successful, which biases the results. Often, it makes sense to distribute samples across known *environmental gradients* (e.g., across soil types, in both *riffles* and pools in rivers, or along elevation gradients). In this case, the environmental variable of interest is mapped, and sampling locations are then distributed within different categories along the gradient.

In general, given the high variance typical in most natural systems, the more samples there are, the better site conditions can be represented, but the number of samples taken will always need to be balanced with the resources available. The higher the variation of a parameter, the more samples are needed to compare the restored and reference systems. The most rigorous way to decide on the number of samples is to collect pilot data and determine the variance, as well as the difference from the objective that one wants to detect (e.g., a less than 5 percent difference in plant cover from the reference system), and then consult with someone who has statistical training to determine the number of samples needed to detect this difference. Elzinga, Salzer, and Willoughby (1998) provide an excellent discussion of how to determine the distribution, size, and number of samples.

Ensuring Quality Control

Monitoring data are collected by multiple people, either paid or volunteer, over multiple years to evaluate whether ecosystems are recovering. To ensure consistent measurements, it is critical to document monitoring protocols and site locations thoroughly. It is not uncommon for entire methodologies to go undocumented, and minor notes about modifications or assumptions are even less likely to be noted, which makes comparisons between samples difficult. Moreover, changes in landform and vegetation can make it difficult to find the original monitoring sites in subsequent years. Monitoring locations should be recorded with GPS to facilitate relocating sampling plots in subsequent years, even if markers are destroyed or removed; for a host of reasons (e.g., burrowing mammals, landslides, actions of people not affiliated with the project), it is common for plot markers to move or vanish entirely from a field site in just a year's time. Even straightforward qualitative monitoring methods, such as photo points, should be carefully documented so that the exact location, direction, and camera frame are recorded.

To compare monitoring data across sites and sampling periods, it is essential to ensure consistent quality across the data sets, particularly when many people, including volunteers and different teams of workers, are involved in the monitoring. Quality control includes ensuring that those doing the monitoring receive standardized training before each measurement period, that the measurement procedures of newly trained individuals are double-checked by and compared with those of experienced observers, and that all data are checked by an experienced individual to detect as many errors as possible.

Data Analysis and Sharing Results

A lengthy discussion of the statistics used in comparing restored areas with either historical data or a reference system is beyond the scope of this book, but numerous other publications address this topic in detail (e.g., Michener 1997; Elzinga, Salzer, and Willoughby 1998; Chapman 1999; Osenberg et al. 2006). It is important to recognize that data entry, proofing, and analysis can be time-consuming undertakings and are seldom budgeted in the costs of monitoring. Consequently, many projects collect extensive data that are never analyzed, making them useless in evaluating whether objectives have been met and informing future management. Hence, monitoring programs and budgets should always address the questions of how, when, and who will analyze the data, and this information should be included in the planning process.

Restoration projects should have a plan for disseminating results with stakeholders, as well as people involved in other similar projects. Sharing both successes and failures from restoration efforts through both formal and informal channels (e.g., talks, field visits, project summaries, formal publications, one-on-one consultations) is invaluable to help improve the success of future projects and minimize "reinventing the wheel." Increasingly, publicly funded projects are required to share data, but implementing this data sharing is still in the early stages. For example, the environmental secretariat for the state of São Paulo in Brazil established a website to report monitoring data on mandatory forest and savanna restoration projects in the state (Atlantic Forest case study; Viani et al. 2017), and a similar database has been set up to share methods and monitoring of prairie restoration projects in the midwestern United States (Walker et al. 2018). Although results are less likely to be shared when projects fail to meet objectives, the lessons learned in these cases should be shared widely to minimize future failures and to improve the cost-effectiveness of forthcoming projects.

Recommended Reading

Elzinga, Caryl L., Daniel W. Salzer, and John W. Willoughby. 1998. *Measuring and Monitoring Plant Populations.* Denver, CO: Bureau of Land Management.
    Provides a thorough overview of all aspects of planning and implementing monitoring programs, as well as data management and analysis.

Evans, Kristen, Manuel R. Guariguata, and Pedro H. S. Brancalion. 2018. "Participatory monitoring to connect local and global priorities for forest restoration." *Conservation Biology* 32:525–34.

Provides justification for and many examples of participatory monitoring in restoration.

Lindenmayer, David B., and Gene E. Likens. 2018. *Effective Ecological Monitoring.* 2nd ed. London: Earthscan.

Provides a sound framework and tips for planning and running an effective ecological monitoring program.

5

Applying Ecological Knowledge to Restoration

The practicalities of planning and monitoring *ecological restoration* projects were discussed in chapters 3 and 4. Now I shift to *restoration ecology*, which is defined as "the science upon which the practice [of ecological restoration] is based" (Society for Ecological Restoration Science and Policy Working Group [SER] 2004, 11). One of the reasons many restoration projects fail is a lack of understanding of the ecology of the system. Fortunately, collaborations between restoration scientists and *practitioners* to incorporate scientific experiments and monitoring in restoration projects provide an excellent opportunity to further our understanding of ecological interactions and processes to improve the success of future restoration efforts (chap. 4; Murcia and Aronson 2014; Palmer, Zedler, and Falk 2016). For example, a combination of scientific studies and *reintroduction* efforts have elucidated the complex relationships between *native* and *nonnative invasive* plants and animals that affect the success of restoration of both tree cacti and giant tortoises in the Galapagos Islands (fig. 5.1; Galapagos Tortoise case study). Likewise, the failure of two reintroduced understory herbs to disperse in a forest at the University of Wisconsin Arboretum led to a better understanding of the role of ants as seed dispersers in this system and the importance of restoring this *mutualism* (Woods 1984).

The *reference model* (chap. 3) should be based on a thorough understanding of the ecology of the *ecosystem* being restored and an understanding of the biology of *focal species*, including information about their dispersal

47

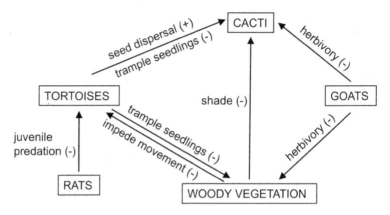

Figure 5.1. Complex interactions between native Galapagos giant tortoises, native tree cacti, woody vegetation, and invasive nonnative rats and goats (Galapagos Tortoise case study). (+) indicates a positive effect and (−) a negative effect of one organism on the other. Removal of an organism has the opposite effect, so removal of goats has a positive effect on both cacti and woody vegetation.

abilities, limitations to establishment, and habitat requirements. The reference model also should consider species interactions, such as *competition*, predation, parasitism, pollination, or seed dispersal, and how these interactions change over time. For example, which species facilitate (enhance) the colonization of a focal species and which compete with it? Is the presence of certain mutualist species necessary for a species of interest to successfully establish? Finally, an understanding of the *cycling* of energy, light, nutrients, and water in a system and how these affect species distributions and *productivity*, is needed.

Rarely does *a natural resource manager* have thorough knowledge of all the various interacting components of an ecosystem, but it is useful to review and synthesize the existing biophysical knowledge about an ecosystem to inform the reference model and guide restoration efforts. Doing so helps identify key actions to improve restoration success (e.g., increasing water flow, introducing a bacterial or fungal mutualist), as well as identify knowledge gaps. The process of researching, discussing, and drawing out visual representations of the ecosystems helps clarify what is known about the system, where more information is needed, and what factors are most important to address to achieve desired restoration outcomes. This information can be synthesized in various ways: from a text summary to an illustration of the distribution of vegetation types as a function of abiotic conditions and site age (fig. 5.2) to more detailed models illustrating key abiotic and biotic interactions affecting focal species (see fig. 5.1).

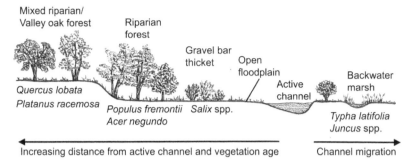

Figure 5.2. Conceptual model of the general distribution of riparian vegetation types in central California (Sacramento River case study) as a function of distance to active channel and successional stage. Modified from Greco (1999). Over time, the right bank will erode during high flow events and the channel will move to the right, exposing new open floodplain habitat that will slowly transition from early successional stages of gravel bar thicket to midsuccessional riparian forest to later successional mixed riparian/valley oak forest. As old channels fill in over time, marsh habitat is created. Drawing by M. Pastor.

Reference models for specific ecosystems or *habitats* should draw on broader ecological frameworks or theories (Palmer, Zedler, and Falk 2016). Ecological theories synthesize past data from many different sites and ecosystems and therefore can be helpful in predicting and guiding restoration outcomes (Török and Helm 2017). The chapters in Palmer, Zedler, and Falk (2016) thoroughly discuss a host of ecological theories and their applications to the science and practice of restoration. In this chapter, I focus on (1) disturbance regimes and models of ecosystem *recovery* and (2) the effect of large-scale spatial processes, such as dispersal of plants and animals, on recovery. I provide a short review of several ecological concepts and highlight how they can guide restoration efforts to improve long-term success. I strongly recommend that readers take a course in or read introductory books on ecology and conservation biology, which cover these topics in more detail. It is equally important to review available information on the ecology of the focal ecosystem or species being restored.

Disturbance Regimes

Most ecosystems are adapted to some form of natural *disturbance* (table 5.1). Natural disturbances are recurring events that change *ecosystem structure, species composition,* or *ecosystem processes.* Natural disturbances are frequently important in maintaining the full suite of species in a given ecosystem type. For example, episodic intense rain and floods control plant

Table 5.1. Restoration of Ecosystems Adapted to Disturbances

Disturbance type	Ecosystem	Adaptation examples	Strategy to simulate adaptation
Fire	Some temperate grasslands, shrublands, forests	Seed germination stimulated by fire, resprouting	Expose seeds to high temperatures or chemicals in smoke to enhance germination
Flood	Wetlands, riparian forests	Plants that withstand extended flooding and grow roots quickly to reach water table	Grow seedlings in tall pots to encourage root growth
Hurricane	Coastal tropical forest	High ability to resprout from broken trunks	Propagate tree species vegetatively through cuttings
Drought	Dryland forests, arid shrublands and deserts	Have deep roots and lose leaves during drought	Grow seedlings in tall pots to encourage root growth
Wind	Dunes	Seed coats that need to be *scarified* by moving sand	Use acid or sandpaper to scarify seed coat
Grazing by large mammals	Savannas and grasslands	Species with basal leaves, species that resprout	Mow to reduce competition from taller-stature species

establishment in *floodplains* and deserts. A number of terrestrial ecosystems have evolved with periodic fires; species such as wiregrass (*Aristida stricta*) and table mountain pine (*Pinus pungens*) in the southeastern United States require fire for reproduction (Vogl 1980). Many tropical forest tree species in coastal regions have adapted to resprout soon after their primary stems are damaged, enabling the forests to recover quickly following hurricanes (Vandermeer et al. 2000).

Disturbance regimes (e.g., type, frequency, magnitude, and timing of disturbance) vary along several gradients that affect recovery, including spatial extent, intensity, predictability, and frequency. For instance, chaparral (shrubland) ecosystems in coastal California evolved with lightning-ignited fires, which likely occurred every fifty to one hundred years (Keeley 2002). In contrast, grasslands in the midwestern United States probably burned every few years in the past, but at lower temperatures due to less plant biomass (Axelrod 1985). Coastal wetlands flood predictably on a diurnal cycle with the tides, whereas many freshwater wetlands flood seasonally, with the intensity, timing, and duration depending on precipitation.

Human actions often alter the frequency and intensity of disturbances. For example, Native Americans decreased the intensity of grazing by hunting large grazing animals and increased fire frequency by intentionally burning *habitats* to favor their desired food plants. More recently, widespread fire suppression in many North American wildlands, aimed at reducing risks to human infrastructure, has resulted in higher fuel loads, which in combination with increasing temperatures and extreme droughts have caused more intense and hotter fires than in the past (Brotons et al. 2013). Likewise, *invasive* woody species introduced by human activities can increase biomass and thereby increase the intensity of fires.

Restoration Implications

Recognizing the disturbance regime with which ecosystems have evolved and how human activities have altered disturbance regimes is integral to designing a restoration project. For example, in arid and semiarid ecosystems where lightning storms ignite fires, restoring a fire regime or doing controlled burns is a common restoration approach. In contrast, wet tropical forests are not adapted to fires, so anthropogenic fires kill most tree species and favor invasive grasses that come from fire-adapted ecosystems. Therefore, wet tropical forest restoration efforts aim to prevent rather than restore a fire regime.

In some cases, changes to the historic disturbance regime are a main cause of habitat degradation, in which case reinstating the disturbance regime is often the most effective way to restore the ecosystem. For example, on the Cosumnes River, an undammed river in central California, initial floodplain forest restoration efforts focused on planting native trees, but these efforts were resource-intensive and plantings had low survival and growth rates (Swenson, Whitener, and Eaton 2003). Following the accidental breaching of a *levee*, extensive seed dispersal and seedling establishment during the high water conditions resulted in the quick establishment of a diverse floodplain forest. Since this accident, resource managers have managed levees and intentionally allowed periodic flooding to restore native vegetation rather than planting trees.

Whereas reinstating the historic disturbance regime can be the most effective way to restore disturbance-adapted ecosystems, it is often not possible given past engineering interventions and ongoing human uses. For example, many rivers are constrained by dams and levees, preventing natural flow and sedimentation regimes. Likewise, fires can threaten human infrastructure and cause air quality problems, making them socially unacceptable in many locations. In these cases, restoration practitioners will need to think about how to simulate disturbances to which ecosystems have evolved (see table 5.1). Repeated experiments have tested the timing and efficacy of controlled releases of water from Glen Canyon Dam to restore the beaches and certain focal species on the lower reaches of the Colorado River (Melis, Korman, and Kennedy 2012). To restore fire-adapted plant species in the absence of a burn, it may be necessary to stimulate germination by exposing seeds to high temperatures or the chemicals present in smoke (Keeley and Fotheringham 1998). Because natural disturbances have many effects, efforts to mimic disturbance rarely result in full recovery of the system, but such efforts can enhance the success of restoration efforts.

Recovery from Disturbance

After a natural or human-caused disturbance, ecosystems go through a gradual change in *abiotic* conditions, *community composition*, and *ecosystem structure* over time in a process known as *succession*. The classical model characterized succession as a predictable progression of communities from an initial, recently disturbed state to a stable, permanent climax (mature) state (Clements 1916). In this model, early colonizing species have high reproductive output, good dispersal abilities, and adaptations to the high light and temperature conditions typical of disturbed areas. Over time, these early colonizing species *facilitate* (make conditions more favorable for) the establishment of later colonizing species, which tend to be dispersal limited, invest more resources in fewer offspring, have more specialized habitat requirements, and be better competitors for limited nutrients, light, and water. This model of ecological succession is consistent with the initial models of ecosystem recovery in the restoration literature that aimed to restore ecosystems to a narrow endpoint after a certain period of time (see fig. 2.1; Bradshaw 1984). This linear model of succession fits the forested ecosystems of the eastern United States fairly well, because it was originally based on that ecosystem, but a host of authors have highlighted numerous ways that this model does not fit the successional *trajectories* of many other ecosystems.

One critique of this linear model of succession is that, as discussed above, many ecosystems do not proceed progressively toward a so-called climax system, but instead experience frequent disturbances that maintain the ecosystem in a dynamic state. For example, flood events in rivers can cause bank erosion and cause the river channel to change course (chap. 6), which exposes open sandbars where early successional vegetation colonizes, creating a mosaic of vegetation types (see fig. 5.1).

Second, species that colonize quickly following disturbance may not facilitate the establishment of later successional species. In ecosystems with strong abiotic limitations like deserts, the species that establish initially often dominate the system over the long term rather than being replaced by successive waves of colonizing plants. Moreover, scientists have long observed that even in ecosystems that experience a predictable progression of species, the species that initially colonize or are planted can strongly influence the successional *trajectory* of the system (Walker and Del Moral 2003; Temperton et al. 2016). In some cases, highly aggressive species, which are frequently nonnative, establish early in the recovery process and impede the colonization of later successional species, thereby slowing the rate of succession. For example, early successional shrubs may prevent the establishment of tropical forest tree seedlings in former pasture lands (Zahawi and Augspurger 1999).

Third, many studies demonstrate that some ecosystems have multiple possible *alternative states* rather than a single climax community (Suding, Gross, and Houseman 2004; Hobbs and Suding 2009). The ecosystem trajectories toward these different states or endpoints can be affected by many factors, including the first species to colonize, abiotic conditions, the intensity of prior habitat degradation, and the species pool in the surrounding landscape that could potentially colonize the site over time (Funk et al. 2008). Grassland-savanna ecosystems are a prime example of alternative ecosystem states ranging from grassland to shrubland to open forest (Briske, Fuhlendorf, and Smeins 2005; Sankaran and Anderson 2009). Historically, grassland ecosystems in the midwestern United States transitioned between shortgrass, tallgrass, and states with higher woody plant cover depending on the frequency and intensity of the grazing and fire (fig. 5.3), creating a mosaic of ecosystem types. Severe anthropogenic disturbances, such as tilling for agriculture, overgrazing by domestic livestock, or clear-cutting of trees, have led to soil compaction and erosion in some areas, a state of transition that is much more difficult to reverse through managing disturbance regimes or active restoration (see fig. 5.3).

These observations of differing trajectories and alternative states have led ecologists to try to develop *assembly rules*, or ways to predict ecosystem trajectories, based on a site's abiotic conditions, the species that establish initially, and the species in the surrounding landscape that might colonize over time (Funk et al. 2008; Temperton et al. 2016). For example, Collinge, Ray, and Gerhardt (2011) conducted an experimental vernal pool (small, seasonally flooded freshwater wetlands) restoration in California in which they manipulated the seeding density and the order in which species were

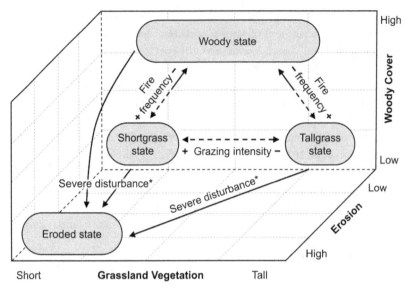

Figure 5.3. Conceptual model of alternative states in grassland ecosystems in the midwestern United States. Dashed lines with arrows represent transitions that are reversible by managing the disturbance regime, and solid lines with arrows represent transitions that are difficult to reverse. Note that as woody cover becomes increasingly dense, shifting back to a grassland state becomes more difficult. *Severe disturbances include tilling for agriculture, overgrazing, catastrophic fires, and overharvesting of woody resources. Figure by J. Lesage. Modified from Briske, Fuhlendorf, and Smeins (2005).

planted. The order in which species were planted influenced community composition over the short term, but vernal pool depth, an abiotic factor, played a stronger role in determining long-term community composition (Collinge, Ray, and Gerhardt 2011).

Restoration Implications

Scientific studies and observations of recovery in restoration projects make it clear that (1) the recovery process is rarely as predictable as suggested by simple, linear successional models and (2) the shape of the successional trajectory varies greatly among ecosystem types and even across individual sites in a given ecosystem (Suding et al. 2016). Nonetheless, it is critical that practitioners gather information, from both observations and prior studies, on the rate and direction of possible successional trajectories as an initial step when planning a restoration project. Knowledge of possible outcomes, and abiotic and biotic factors that are likely to affect these outcomes, should guide actions to aid the recovery process.

For cases in which ecosystem recovery is rapid and the successional trajectory is consistent with restoration *goals* and *objectives* (Gann et al. 2019), *natural regeneration* is often the least intensive and most cost-effective restoration approach (Prach and del Moral 2015; Chazdon and Guariguata 2016). In such cases, actively intervening to restore the system may be a waste of resources or may even slow recovery. For example, pastures in central Brazil show high regeneration of resprouting trees. Planting tree seedlings to restore savanna is resource intensive and damages resprouting trees, so there is no net gain in tree stem density (Sampaio, Holl, and Scariot 2007).

A wise strategy, if socially feasible, is to wait a few years before actively intervening in restoration, during which time practitioners can assess the rate and species composition of natural regeneration (Murcia and Aronson 2014). If a site recovers at an intermediate rate, then a good approach may be to assist regeneration (chap. 2); typical *assisted regeneration* actions vary by ecosystem type, but may include restoring ecological flows and fish passage in aquatic systems, removing competitive *invasive species* that outcompete native species, or clearing firebreaks to reduce fire risk in non-fire-adapted systems (Shono, Cadaweng, and Durst 2007; Gann et al. 2019).

In ecosystems that are slow to recover, it is important to characterize the successional model and identify the factors limiting recovery before carefully selecting *active restoration* strategies. Is the ecosystem so degraded that it is necessary to completely reconstruct the *topography*, *hydrology*, or other *abiotic* conditions to provide the appropriate habitat conditions for the native ecosystem and focal species to recover? If focal species do not colonize naturally, the next step is to actively reintroduce those species.

Practitioners must also recognize that the establishment of a specific suite of species at the outset of a project does not guarantee that additional species will colonize naturally, an assumption that has been referred to as the "field of dreams" (Hilderbrand, Watts, and Randle 2005). Even when desired species colonize, ongoing monitoring and *adaptive management* (chap. 4) are necessary to ensure that a restoration project follows the desired successional trajectory. Restoration projects often require ongoing *management*, such as removing invasive species and introducing later successional species that do not readily colonize.

Ecosystem succession following both natural and anthropogenic disturbances is a long-term process that can take many years, decades, or centuries, even with human management to accelerate the recovery process, but there is strong pressure to meet project objectives and demonstrate favorable outcomes within a few years. The result is a tension between short- and long-term project goals and resources. For instance, preventing erosion by establishing ground cover quickly may be an important goal in a highly disturbed system, but many studies have shown that planting aggressive nonnative species for erosion control inhibits the establishment and growth of later successional species (Holl 2002b). Restoration project budgets typically span one to a few years rather than the appropriate length of time needed to manage and guide the long-term successional trajectory of an ecosystem. There is no easy answer to balance this temporal mismatch. In the best-case scenario, targeted objectives will be set that correspond with sequential reference models that represent stages along the recovery trajectory (Clewell and Aronson 2013). Then monitoring will evaluate whether intermediate objectives are being met that are consistent with longer-term ecosystem recovery, and, if not the case, corrective actions will be taken (chap. 4).

Ecological Processes at Large Spatial Scales

Viewing restoration projects at a large spatial scale is essential because human activities often affect ecosystems far beyond their visible boundaries and because the recovery of disturbed ecosystems is strongly influenced by physical and ecological processes in areas surrounding the restoration site (Holl, Crone, and Schultz 2003; Metzger and Brancalion 2016). Table 5.2 lists several such large-scale processes that affect *riparian* forest recovery and restoration, such as flooding, water flow rate, nutrient inputs, and movements of plants and animals.

The terms *landscape* and *large-scale* are variably defined. Forman and Godron (1981, 733) describe a landscape as a "cluster of interacting ecosystems" that exchange organisms and materials (e.g., water, nutrients) across "a kilometers-wide area," whereas Metzger and Brancalion (2016, 91) define a landscape as a "heterogeneous mosaic composed of interacting landscape units" and emphasize that the scale of the landscape depends on the range of the focal organism or process. Some herbaceous plants respond to the small-scale (less than 1- to 5-meter) distribution of water and nutrients, and their seeds only disperse a few meters to tens of meters to colonize new sites, yet they may be grazed by insects or mammals that move many kilometers and be influenced by nitrogen deposition from human activities tens of kilometers away or more. Clearly, restoration practitioners should consider how the surrounding areas affect the site being restored, and the spatial scale to consider will be defined by the relevant ecosystem, organisms, or processes.

In this section, I discuss theories about habitat patch size and landscape

Table 5.2. Ecological Processes Operating at Large Spatial Scales That Influence Recovery and Restoration of Riparian Forests

Physical processes
- Water flow rate
- Water drawdown rate
- Flooding (frequency, timing, duration, magnitude)
- Scouring and erosion
- Sediment and nutrient deposition
- Chemical movement (fertilizers, pesticides)

Population processes
- Dispersal and colonization of seeds
- Gene flow (seeds and pollen)

Community processes
- Movement of seed dispersers, pollinators, and other mutualists (e.g., *mycorrhizae*)
- Movement of herbivores, seed predators, and parasites
- Dispersal and colonization of invasive species

Human alterations to processes
- Dams
- Levees
- Groundwater pumping
- Land use changes (e.g., conversion of land uses, farming practices)
- Precipitation (climate change)

Source: Modified from Holl, Crone, and Schultz 2003.

connectivity and their implications for allocating restoration efforts at a large scale. Restoration of physical processes, such as hydrology and nutrient flows at a range of spatial scales, is discussed in chapter 6.

Habitat Patch Size

Larger areas or patches of habitat are desirable for several reasons. First, larger areas host more species than smaller patches if the habitat is of similar quality (MacArthur and Wilson 1967; Ewers and Didham 2006), in part because larger habitat patches typically have a wider range of resources for varied species. Moreover, larger patches have enough space to host species that require big home ranges, such as some birds and mammals.

Second, large habitat patches usually host larger populations of individual species than smaller patches, and larger populations are less prone to *extinction* (Lande 1993; Metzger and Brancalion 2016). Small populations are often highly susceptible to extinction due to *environmental stochasticity* (random variation in climatic conditions or natural disturbances) and natural variability in birth and death rates. For example, a rare butterfly species with only one or a few populations is at higher risk of extinction from extreme events that result in a dramatic decline in host plant populations than is a species with a sizable population. Likewise, small populations tend to have low genetic variation, which can lead to *inbreeding depression*, the process by which detrimental genes accumulate in offspring. Many big cat species, such as the Florida panther (*Puma concolor*), now have such small populations that they suffer from inbreeding depression, and many of their offspring are infertile.

Third, habitat borders or edges, especially in the case of forests, often experience *edge effects*, where abiotic conditions are more extreme (Murcia 1995). Think about when you walk along a forest edge: it is warmer, drier, and brighter than farther inside the forest. Edge effects can range from a few meters to more than a hundred meters into a forest, depending on the organism, the variable measured, and the contrast between the two habitats (Cadenasso et al. 2003). Abiotic conditions at habitat edges tend to favor disturbance-adapted, generalist species that outcompete species that require intact, interior habitat. For these reasons, invasive nonnative species tend to be more common at the edges of forests. For example, invasive vines, such as kudzu (*Pueraria* spp.) in the southeastern United States, rapidly grow over native trees and shrubs, particularly at forest edges. Because larger habitat patches, and those with a more regular shape, have a lower ratio of edge habitat to overall patch area (Metzger and Brancalion 2016), they host more species.

Finally, it is more feasible to restore ecosystem processes (such as flows of water and nutrients) and disturbance regimes in larger habitat patches than in smaller ones. In situations like the Younger Lagoon case study, where small patches of grassland habitat are being restored in an urban landscape, it simply is not possible to restore a historic fire regime given concerns about the fire spreading and causing damage to nearby homes. Likewise, reintroduction of large grazers is not practical in small habitat patches.

Although large habitat patches are desirable for many reasons, some small habitat patches harbor unique species or unusual habitat characteristics (e.g., rare soil types), which makes them important for conservation and restoration (Simberloff and Abele 1976). Therefore, not only the size but also the quality of the habitat being restored must be considered. Nonetheless, restoration efforts should prioritize larger areas of habitat or areas adjacent to existing remnant habitat to enlarge remnant patches when possible (fig. 5.4A). Doing so may serve to reduce edge effects in remnant habitat over time, and the remnant habitat will provide a source of colonizing plants and animals for the restored area.

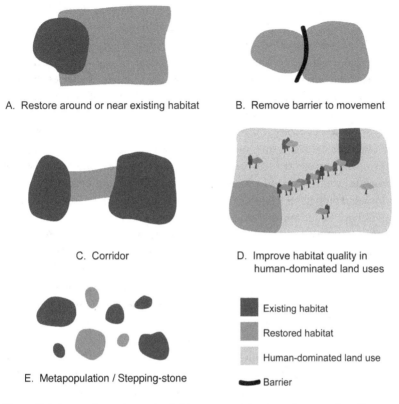

Figure 5.4. Approaches to increasing habitat connectivity through restoration. Figure by A. Calle.

Increasing Landscape Connectivity

It is desirable to conserve and restore large areas of habitat, but given the spatial scale of human activities, doing so is only realistic in a limited number of situations. More often, restoration projects are embedded within a human-dominated landscape. Many large carnivores, forest-dependent birds and the plant species they disperse, and other species with large home ranges are increasingly isolated to remnant habitat patches because of the difficulty of moving through human-dominated land uses. Hence, restoration efforts should consider how to increase *landscape connectivity* to best facilitate the movement of flora and fauna and to enhance natural colonization in restoration sites (Metzger and Brancalion 2016).

All models of succession assume that different species will colonize a patch over time, which is largely dependent on source populations nearby.

Many grassland plant species do not have long-distance dispersal mechanisms (Seabloom et al. 2003), and in tropical forests, few animal-dispersed tree seeds are dispersed more than 10 to 100 meters beyond existing forest edge (Holl 2012). Therefore, restoring areas near source populations is desirable whenever possible to enhance the number of species that can colonize naturally, reducing the overall cost and effort of actively reintroducing species (see fig. 5.4A).

In addition, ecological restoration should aim to remove barriers to movement of organisms, nutrients, and water (fig. 5.4B). Many aquatic restoration projects focus on removing dams, levees, and roads, which impede water flow and therefore the colonization and establishment of flora and fauna. For example, more than sixteen hundred dams of various sizes have been removed over the past century in the United States alone (American Rivers n.d.), which allows for the transport of nutrient and sediment downstream and the movement of various aquatic organisms in both directions (chap. 6). In places where roads divide habitat patches, wildlife underpasses (tunnels) for ground-dwelling animals or overpasses (bridges or rope ladders) for tree-dwelling animals have been used with some success to facilitate animal movement (see book website for illustrations of some structures to enhance faunal movement).

One approach to enhancing the movement of organisms among habitat patches is to restore *ecological corridors* that link existing patches (fig. 5.4C). Corridors can increase movement among existing habitat patches and hence serve to increase *gene flow* among populations, reducing some of the problems of small populations discussed earlier, such as inbreeding depression. In Queensland, Australia, several rain forest corridors, roughly 100 meters wide and up to 1.2 kilometers long, have been planted with a diverse suite of native tree species, linking remnant forest patches (Tucker and Simmons 2009). The planted trees closed canopy within a few years, which has facilitated the establishment of a diversity of additional rain forest plants within the corridor and increased the movement of small mammals between habitat patches.

Although ecological corridors can improve landscape connectivity, their usefulness depends on the width of the corridor and the biological needs of the focal species. Whereas small mammals might use a 20-meter-wide corridor, some tropical birds and large mammals need corridors that are 200 meters wide or more (Lees and Peres 2008). In addition, the efficacy of corridors depends on the quality of habitat they provide. Because of their shape, corridors have a high ratio of edge to interior habitat, typically making them low-quality habitat for many species with specialized habitat needs. At the same time, this high edge ratio makes them valuable sources for dispersion of seeds favoring natural regeneration in adjacent abandoned lands (Rey Benayas and Bullock 2015).

Natural resource managers can increase movement among remnant and restored habitat patches by managing the human-dominated land uses in between these habitats to promote movement of fauna (fig. 5.4D). Within agricultural landscapes, it is typical to restore *buffer strips* (narrow strips along the edges of fields that are adjacent wetlands or rivers) to capture sediments and improve water quality. Buffer strips and *hedgerows* (lines of closely planted shrubs or trees at the edge agricultural fields) serve as narrow corridors to facilitate the movement of some species across the landscape (Rey Benayas and Bullock 2015). Moreover, in originally forested landscapes, increasing tree cover within pasture and croplands can facilitate the movement of a range of fauna, including seed-dispersing birds and bats (Mendenhall et al. 2011). Reducing pesticide use in agricultural lands improves habitat quality for insects and other fauna that eat them. Although only a subset of species will leave remnant habitat and move through a human-dominated landscape, improving the habitat quality of human land uses increases the number of species that will do so.

Whereas restoring corridors and managing agricultural lands to improve overall habitat value will enhance the movement of some species, other species are naturally distributed patchily in *metapopulations*, sets of geographically isolated subpopulations interconnected by colonization and gene flow between subpopulations. A metapopulation often consists of one or more larger populations that are persistent and several smaller populations that may go extinct when conditions are unfavorable but are recolonized periodically when conditions improve. These subpopulations fluctuate separately, with occasional movement of organisms among them that serves to recolonize available habitat patches and redistribute genetic material. Species that rely on naturally patchily distributed habitats often have a metapopulation structure. Examples of organisms with metapopulations are amphibians that live in ponds, insects that rely on host plants that are distributed on patchily distributed soil types, and many marine organisms (e.g., reef fish, oysters, lobsters) that live in separate subpopulations. For these species, restoring habitat patches that can serve as stepping-stones between populations may help enhance the regional persistence of the species (fig. 5.4E). For example, McIntire, Schultz, and Crone (2007) have collected extensive data on habitat needs and movement behaviors of the endangered Fender's blue butterfly (*Icarus icaoides*), which relies on small wetland patches in the northwestern United States. They have incorporated this information into large-scale models to help land management agencies prioritize conservation and restoration of available wetland patches to best connect some of the larger subpopulations.

Restoration Implications

Restoration projects should be designed in the context of the surrounding land uses with an understanding of how both local and large-scale processes affect the restoration site and focal species. Planning for larger projects should include considerations of how to increase connectivity most effectively to restore physical processes, colonization of different species, and a complex suite of species interactions to facilitate natural successional processes. For small projects embedded in urban areas or intensively farmed agricultural landscapes, it is challenging to restore physical processes and successional dynamics. Therefore, ongoing active manipulation of the habitat is usually necessary to restore the system to a state that resembles the reference model.

The most suitable way to allocate restoration efforts across a large area will depend on the ecology of the focal species, as well as the surrounding land uses and the project constraints (Metzger and Brancalion 2016). For example, deciding whether it is more effective to restore corridors or small patches as stepping-stones depends on the biology and historical distribution of the focal species and ecosystems, as well as the availability of land for restoration. In most cases, choices about where and how to restore are constrained by land ownership, competing land uses, and cost, so practitioners must work with what is available. For example, large-scale forest restoration projects are more likely to succeed on lands that are not productive for agriculture (Latawiec et al. 2015), where they will protect water supplies used by downstream communities, or in areas that have been set aside for conservation purposes than in other areas. In highly productive agricultural lands, it is more feasible to integrate forest restoration within a mosaic of other human-dominated land uses by increasing tree cover in the agricultural landscape to increase connectivity than by other methods. Silvopastoral systems in Colombia provide an example of where this approach has been successful in meeting both ecological and human needs at the landscape scale (Calle et al. 2013). There, *nitrogen-fixing* shrubs and trees are planted both within pastures and in hedgerows at the edge of fields to provide improved fodder for cattle and shade, which improves milk production. At the same time, these changes have increased the number of bird species and reduced soil erosion. Regardless of the specific constraints of a site, it is important to consider the reference ecosystem model and the spatial context of the restoration project when deciding how to allocate resources most effectively.

Recommended Reading

Metzger, Jean Paul, and Pedro H. S. Brancalion. 2016. "Landscape ecology and restoration processes." In *Foundations of Restoration Ecology*, 2nd ed., edited by M. A. Palmer, J. B. Zedler, and D. A. Falk, 90–120. Washington, DC: Island Press.
Provides a good overview of large-scale ecological processes in restoration ecology.

Palmer, Margaret A., Joy B. Zedler, and Donald A. Falk. 2016. *Foundation of Restoration Ecology*. 2nd ed. Washington, DC: Island Press.
Includes numerous chapters about ecological concepts and theories underlying restoration ecology in an edited volume.

Suding, Katherine, Erica Spotswood, Dylan Chapple, Erin Beller, and Katherine Gross. 2016. "Ecological dynamics and ecological restoration." In *Foundations of Restoration Ecology*, 2nd ed., edited by M. A. Palmer, J. B. Zedler, and D. A. Falk, 27–56. Washington, DC: Island Press.
Discusses different successional models of ecosystem recovery.

6

Landform and Hydrology

Many human activities completely change the physical conditions of an area. For example, humans build mines and networks of roads in terrestrial *ecosystems*, channelize rivers, and excavate and fill wetlands. Moreover, surrounding land uses strongly affect both water and chemical fluxes in a restoration site. For example, more than fifty thousand large dams block river flows across the globe, altering downstream conditions (International Rivers 2014). In agricultural landscapes, both nutrient and sediment inputs to nearby water bodies are typically elevated due to substantial use of fertilizer and increased erosion of bare soil.

It is critical to restore the *abiotic* conditions (e.g., *topography*, nutrient availability, hydrologic regime, light, soil characteristics) at a site first to enable the recovery of the desired microorganism, plant, and animal communities. For example, if you *revegetate* before correcting the hydrologic flow paths in terrestrial systems where there is extensive gully erosion, then the vegetation will be likely washed away. If the correct hydroperiod of a wetland is not restored, then it is impossible to restore the plant communities that are adapted to the historical timing and depth of *inundation*.

The dynamics of rivers, wetlands, and lakes are driven by their *hydrologic regimes*, namely the magnitude and timing of the flow of water, which are influenced by both surface and groundwater inputs from throughout the *watershed* (Roni and Beechie 2012). As the National Research Council (1992, 184–85) assessment of aquatic restoration aptly states, "Rivers and their *floodplains* are so intimately linked that they should be understood, managed, and restored as integral parts of a single ecosystem." Nonetheless, Palmer, Hondula, and Koch (2014) reviewed 644 river restoration projects worldwide and found that only 4 percent of them were implemented at the watershed scale.

In this chapter, I discuss challenges to and strategies for restoring the abiotic characteristics of *landform* and hydrologic regimes in terrestrial and aquatic systems. I emphasize the importance of restoring abiotic patterns and processes at both small and large spatial scales. These physical conditions and processes are inextricably linked with soils, water quality, and nutrient *cycling*, which I discuss in chapter 7, and lay the foundation for the successful restoration of an ecosystem's biota, discussed in chapters 9 and 10.

Human actions, such as leveling lands and filling wetlands for agriculture, often homogenize both the abiotic and *biotic* characteristics of ecosystems. Unfortunately, it is common to apply homogeneous restoration strategies across large areas, which leads to less diverse ecosystems. Abiotic conditions often vary over the space of a few meters, leading to small-scale *spatial heterogeneity* in natural ecosystems. This uneven distribution of resources, combined with variable dispersal patterns and species interactions, leads to patchy distributions and coexistence of different species (Larkin, Bruland, and Zedler 2016). Restoration strategies should be designed to re-create the natural spatial heterogeneity of the ecosystem, such as variable water depths in wetlands or concentrating soil nutrients in patches in shrublands (chap. 7). This approach not only more closely reflects the historical *reference model*, but also increases the chance of restoration success under variable climatic conditions. For example, Doherty and Zedler (2015) found that creating small-scale heterogeneity in elevation (ranging by 10 to 40 centimeters) in a wet meadow restoration project in Wisconsin resulted in varying soil moisture, which in turn resulted in higher growth and survival of *native* plants in higher-elevation microsites in a dry year and better outcomes in lower-elevation sites in a wet year.

Where humans have dramatically altered the landform and water movement, restoration may require substantial earth-moving and heavy machinery at large scales to recontour topography, restore river channel patterns, or remove dams or other barriers to water flow. These efforts require detailed maps of topography, soil type, and vegetation cover and may involve simulations of water flow patterns to predict how specific restoration actions will affect the hydrologic regime. In places where prior *degradation* is less intense or projects are smaller, less intrusive approaches may serve to modify topographic patterns or water flow paths, such as installing erosion control cloth, planting vegetation, or introducing logs or stones to redirect river flow.

Terrestrial Landform and Hydrology

In natural systems, some precipitation (rainfall and snowmelt) infiltrates into the soil and ultimately may percolate into the groundwater system. Some is taken up and transpired by the plants. The water that exceeds the infiltration capacity of the soil will run off the ground surface and directly contribute to streamflow (fig. 6.1A). This surface runoff is slowed by vegetation and other features that increase surface roughness, such as mulch or terraces. When heavy machinery has altered the site morphology, more than just the slope of the site is affected. Heavy machinery can remove vegetation and increase *soil compaction*, reducing water infiltration into the soil (Whisenant 1999). The combined changes in slope, soil compaction, and vegetation cover affect water runoff patterns and microclimate (Roni and Beechie 2012). First, runoff increases, causing gullies and soil erosion (fig. 6.1B), which negatively affects water quality. Second, the changes in slope and clearing of vegetation result in more extreme temperature and light conditions in the area, affecting *habitat* suitability for plants and animals. Third, water availability can decrease or increase. When land has been leveled, soil compacted, or water flow blocked, water may pond and flood plants. In places where precipitation quickly runs off steep slopes with compacted soils rather than infiltrating into the soil, plants are more likely to experience drought conditions during dry periods.

Restoration Strategies

The first step to reducing erosion and restoring water drainage patterns in highly degraded sites is to restore the *topography*, meaning the shape of the land. Indeed, legislation in many countries requires companies to restore the approximate original contour of the land and revegetate it following mining and construction (chap. 11). This process starts with developing a reference model for topography and water flows from similar landforms in the region and numerical modeling to minimize erosion and ensure slope stability at the restored site (Bugosh and Epp 2019). Then, heavy machinery is used to move soil to create a specific slope grade, length, and aspect; remove roads; or level gullies. Efforts to restore topography and runoff patterns are more successful when the original soil layers are replaced (chap. 7) and soil compaction is reduced to increase water infiltration. In some cases, simply restoring topography and water flow patterns is sufficient to catalyze recovery. For example, in Redwood National Park

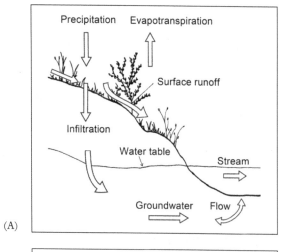

(A)

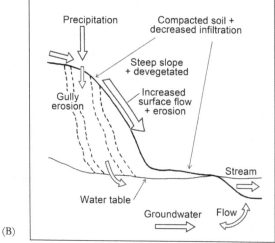

(B)

Figure 6.1. Hydrologic flow paths in (A) a natural system and (B) a disturbed system with altered topography, compacted soils, and minimal vegetation. Evapotranspiration is the transfer of water from land to the atmosphere both through evaporation from the plants and soil and transpiration from plants. Drawings by M. Pastor.

in northern California, approximately $10 million was spent between 1977 and 1990 to remove more than 450 kilometers of old roads to reduce erosion into nearby streams (Steensen and Spreiter 1992). Mulch was used to temporarily minimize erosion following earth-moving, and natural recovery of vegetation occurred so quickly that it was not necessary to plant or seed vegetation. More commonly, after restoring the site's topography, various techniques are used to reduce erosion, including actively revegetating the area.

Soil erosion is a function of the soil compaction, amount of water runoff, slope length and steepness, vegetation cover, and land *management* practices employed. Hence, various management practices are used to increase infiltration, slow and redirect water flow, and reduce erosion, which in turn enhance the survival of naturally establishing and planted vegetation. These practices are used after recontouring the land in highly disturbed sites, as well as in sites with less severe erosion and water flow problems that have not had extensive earth-moving. Restoration *practitioners* employ different ground covers, such as erosion control cloth, hay mulch, wood mulch, or compost, to reduce surface soil erosion from both wind and water, retain surface soil moisture, and reduce soil temperature (Bradshaw and Chadwick 1980; Munshower 1994). However, care must be taken that seeds of undesired species are not introduced with any plant residues or hays that are applied to a site.

Several strategies serve to reduce or redirect water flow. Small rolls of erosion control cloth or logs placed along steep slopes break the flow of water and trap sediments (fig. 6.2A). Likewise, *waterbars*, comprising logs, rocks, or raised soil, are often built diagonally across trails or roads to slow and redirect water flow to minimize gully formation (fig. 6.2B). Rocks or sandbags placed in small gullies redirect and slow the water flow, trap sediments, and reduce further erosion.

Establishing vegetation (chap. 9) is an important step to reduce erosion, increase water infiltration into the soil, and improve water quality (chap. 7). Hence, many cities are restoring green spaces, in part to reduce runoff and improve quality downstream in the watershed (Doherty et al. 2014). In bare and desertified arid systems that are subject to wind erosion, windbreaks of shrubs or trees, or even fencing, reduce soil erosion and trap blowing sand.

Finally, a few techniques are used to roughen the soil surface, increase water infiltration and root penetration, and provide sheltered microsites to facilitate seed germination (Whisenant 1999). Highly compacted soils are commonly *ripped* before planting; the surface of soil is broken by hook-shaped tines mounted on the back of a tractor or dragged across the soil surface by hand (Bradshaw and Chadwick 1980; Munshower 1994). This technique has drawbacks, however, as ripping the soil may temporarily increase erosion and facilitate the invasion of weedy species; therefore, it is more suited to flatter lands and should be followed by other strategies to

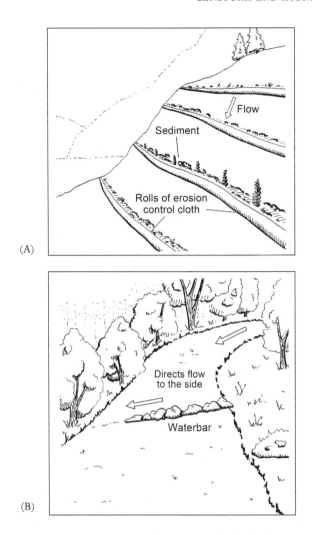

Figure 6.2. Restoration strategies to reduce erosion. (A) Rolls of erosion control cloth serve to slow the flow and trap sediment and seeds. (B) A waterbar is used to direct flow to the side of a trail and minimize gully formation. Drawings by M. Pastor.

control erosion, such as covering the soil surface with mulch. In arid lands, creating *microcatchments* (small depressions in the ground; fig. 6.3) serves to concentrate water, nutrients, and seeds in certain locations, which in turn increases spatial heterogeneity and provides *microsites* that are more favorable for seedling establishment (Whisenant 1999).

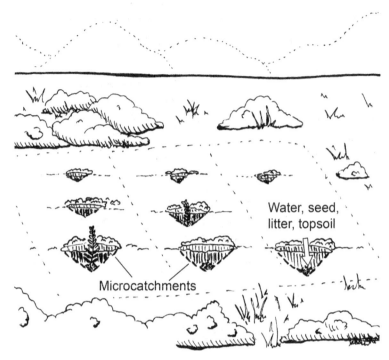

Figure 6.3. Microcatchments used to create small-scale topographic heterogeneity and concentrate water, seeds, topsoil, and litter in arid systems. Drawing by M. Pastor.

Wetland Topography and Hydroperiod

Wetlands have attributes of both terrestrial and aquatic systems. They experience permanent or periodic inundation (i.e., flooding) that dramatically influences their soils and vegetation. The *hydroperiod*, which refers to the depth, duration, frequency, and seasonality of inundation, varies by wetland type (table 6.1). It is affected by precipitation and surface runoff, groundwater inputs, topography, and the *soil texture* (Craft 2016). The hydroperiod drives soil chemistry; wetland soils are low in oxygen, due to slow movement of oxygen through water, and as a result, they often have high *organic matter* content due to slow decomposition. Coastal wetlands also vary along a salinity *environmental gradient* ranging from freshwater to brackish to salty, depending on their tidal cycles and freshwater inputs. Wetland plant species are distributed along a gradient of inundation and salinity, which may vary over small spatial scales (fig. 6.4). Plants that can tolerate flooded wetland soils for extended time periods have evolved

Table 6.1. Abiotic Characteristics of Common[1] Wetland Types

Type	Duration and seasonality of inundation	Salinity	Other characteristics
Mangrove	Changes daily with tides	Saline	Trees, located in tropical areas
Salt marsh	Changes daily with tides	Saline to brackish	Various short to medium-stature plants
Swamp	Permanently inundated	Freshwater	Usually have trees
Riverine / riparian	Varies seasonally with river flow	Freshwater	Habitat is dynamic as river channel meanders
Vernal pool, Prairie pothole	Seasonally inundated (vernal) or permanently wet (prairie)	Freshwater	Primarily forbs, sedges, rushes, and grasses
Bog, Fen, Peatland	Permanently inundated	Freshwater	Extremely high-organic-matter soil, acidic in bogs, calcareous or neutral in fens; may receive inputs from groundwater

[1]Wetland types are distributed along multiple abiotic gradients; many more subtypes are described in Craft 2016.

mechanisms to allow oxygen to reach their roots, such as hollow stems and aerial roots.

Globally, 64 to 71 percent of wetland area has been lost since 1900 (Davidson 2014). Large areas of wetlands have been excavated or filled to create more upland or open-water habitat. Water flow rates generally slow as water passes through wetland vegetation and soil, causing sediments suspended in the water to settle and accumulate in wetlands. Hence, vegetation clearing, agriculture, and other human activities commonly result in increased sediment deposition in both wetlands and lakes, altering the ecology and nutrient cycling of these ecosystem (chap. 7). In contrast, dams trap sediments upstream instead of the sediment flowing downstream and accumulating in wetlands, which can cause the wetlands to sink and the vegetation to die from prolonged inundation. This situation is particularly problematic for coastal wetlands, which are also threatened by rising sea levels. For example, along the coast of Louisiana, 5,000 square kilometers of wetlands have been lost in the past century due to a complex set of drivers, including reduced sediment deposition due to dams and levees, dredging of navigation canals, and sea-level rise (Jankowski, Törnqvist, and Fernandes 2017). Changes to the amount of precipitation

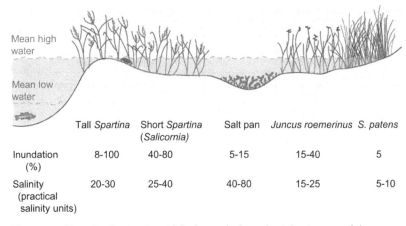

	Tall *Spartina*	Short *Spartina* (*Salicornia*)	Salt pan	*Juncus roemerinus*	*S. patens*
Inundation (%)	8-100	40-80	5-15	15-40	5
Salinity (practical salinity units)	20-30	25-40	40-80	15-25	5-10

Figure 6.4. Plant distribution in a tidal salt marsh along the Atlantic coast of the United States as a function of topography, inundation, and salinity. Figure modified from Craft 2016.

and the increasing intensity of storm events, combined with increasing temperatures, threaten both coastal and noncoastal wetlands (Osland et al. 2016).

Wetland loss is a major concern because wetlands provide important *ecosystem services* such as flood control, *carbon storage*, and protection from coastal erosion and storm surges (Asian Mangrove case study; Galatowitsch and Zedler 2014; Craft 2016). Wetlands also take up nutrients along with sediments, thereby improving water quality and providing important breeding grounds for many fisheries. Because of these services, wetlands are legally protected in some countries (chap. 11) and are the focus of extensive restoration efforts.

Restoration Strategies

Successful wetland restoration depends critically on re-creating a natural hydroperiod (Asian Mangrove case study; Galatowitsch and Zedler 2014). Doing so requires restoring topography and underlying soil layers that affect drainage patterns (discussed here) and restoring hydrologic connectivity, including water inputs from groundwater, surface water runoff, rivers, or tidal influences (discussed in the next section).

As with terrestrial systems, wetland topography is restored by first developing a reference model of the gradient in ground surface elevation needed to restore the desired hydroperiod and associated biotic communities. Then heavy machinery is used either to excavate areas that have been filled or to add sediment where the elevation needs to be increased, after which sediments need time to dewater and settle. Sometimes sediments are transported in large pipes from one location to another (see Marsh Creation video in online resources). In some cases, sediments are left to revegetate naturally, which can happen quickly if there are nearby sources of seeds and plant fragments that disperse via water (Galatowitsch and Zedler 2014). In other situations, sediments are stabilized through active revegetation (chap. 9).

Increasingly, coastal wetland restoration projects are focused on increasing the soil surface elevation so that tidal wetlands do not drown in response to sea-level rise and continue to provide protection against coastal storm surges and erosion. They are part of a larger effort to use vegetation and other natural materials to create "living shorelines" rather than using engineering solutions such as dikes and levees that provide little habitat value and are more costly to install and maintain (Craft 2016; Narayan et al. 2016; Parker and Boyer 2017). Many studies have shown that planting vegetation, establishing oyster or coral reefs, or placing rocks, mounds of oyster shells, or logs in key locations can trap sediment and increase soil surface elevation over time so that the shorelines are dynamic or "living." Moreover, organic matter accumulates in most wetland types, which also increases elevation. Drexler et al. (2019) found that restoring hydrologic connectivity in a salt marsh on the coast of the state of Washington resulted in approximately 5 centimeters of accumulation of sediments in the six years following dike removal.

Re-creating the gradual changes in topography that are characteristic of natural wetlands is challenging but important, as even a small difference in wetland elevation can dramatically affect the duration and depth of flooding and therefore the plant species composition (see fig. 6.4; Collinge, Ray, and Gerhardt 2011; Doherty and Zedler 2015). Wetland restoration efforts often create more open water and upland habitat areas, as well as proportionally smaller areas with fluctuating hydroperiods that many wetland species require to thrive (National Research Council 1992; Craft 2016).

It is important to consider the texture of the underlying soil layers in site selection because they strongly affect nutrient cycling, drainage, and sediment movement (chap. 7; Boyer and Zedler 1998; Craft 2016). Some freshwater wetlands, such as vernal pools and prairie potholes, are formed in locations where an underlying clay layer limits the rate at which water drains from the site. Doherty et al. (2014) compared three freshwater wetlands with similar size, shape, elevation, topography, and soils that were excavated and seeded with prairie herbs to treat stormwater runoff from an urban watershed. They found that differences in the underlying clay layer resulted in up to a fivefold difference in the rate that water drained. The site with slower drainage became dominated by cattails (*Typha* spp.), a widespread, highly competitive herbaceous species, and provided less stormwater retention and nutrient uptake services than areas with better drainage, demonstrating the importance of correct soil layers to restore the hydroperiod and associated wetland ecosystem habitat and functions.

Hydrologic Regimes and Channel Meandering

Rivers are highly dynamic systems, with much of the activity happening during rare peak flow events when the river flow mobilizes large woody debris, rocks, finer sediments, and dissolved nutrients; overflows the channel onto the floodplain; and at times changes the location and shape of the channel. As the flow slows down, suspended sediments drop out of the water and accumulate on the floodplain. A river's flow regime is characterized by the peak and base flows; the frequency, duration, and timing of flooding; and how fast the flow increases and decreases (fig. 6.5). The peak flow after a major precipitation event may be one hundred or one thousand times that during the dry season, and some streams dry up entirely for a portion of the year.

Water flows from throughout the watershed interact with the local geology, topography, and vegetation to affect sediment movement and channel patterns (Roni and Beechie 2012; Palmer, Hondula, and Koch 2014). The greater and faster the flow, the more sediment and larger objects that are mobilized. Some rivers naturally meander (i.e., their channels move laterally over time) throughout their floodplain, creating riparian wetlands and *point bars* where vegetation can establish (fig. 6.6A). Other rivers are braided with multiple interconnecting channels (fig. 6.6B). Rivers have heterogeneous in-stream morphologies, which include shallower fast-flowing water (*riffles*) and deeper and cooler pools that serve as habitat and *refugia* for aquatic organisms (see fig. 6.6A).

Water flows affect riparian vegetation dynamics and fauna, as riverine flora and fauna have adapted to the timing of high and low flow events. For example, many riparian plant species disperse via floating seeds or vegetative parts and establish best during the historical timing of peak flow events. If the water level drops slowly following a flood, then riparian trees and shrubs have time to establish and grow roots that reach the water table (Wohl, Lane, and Wilcox 2015). The streambed substrate, water quality and clarity, temperature, and water depth and velocity, particularly at low flows, all determine that habitat's suitability for riverine fauna (chap. 10).

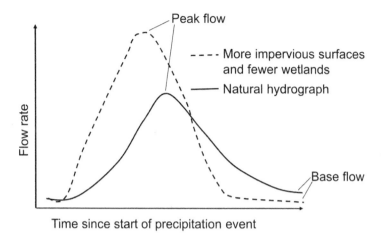

Figure 6.5. Stylized river flow regimes. Solid line shows the rise and decrease of river flow rate following a precipitation event in a natural system. Dashed line illustrates the faster rate of increase and decrease in flow rate and the greater peak flow in systems with more paved surfaces and less wetland area to slowly absorb and release water.

Riparian flora and fauna also influence channel dynamics and in-stream conditions. Roots of riparian vegetation stabilize soil, reduce erosion, and affect channel meandering. Riparian trees provide nutrient inputs through litterfall and reduce stream temperature by shading the channel. In some regions, beavers make dams that slow flows and increase localized inundation and channel patterns.

Humans have altered the hydrologic regimes of most rivers, wetlands, and lakes worldwide through water withdrawal, paving large areas of watersheds, and damming and straightening rivers. Extensive water withdrawals for agriculture, power plant cooling, and domestic uses reduce water inputs to natural systems, which is particularly problematic in arid and semiarid ecosystems. For example, in the Murray-Darling Basin, which covers one-seventh of the Australian continent, twenty of the twenty-three river valleys were rated in poor or very poor health, largely due to extensive withdrawal of water for irrigation (Docker and Robinson 2014). In addition, the extent of paved surfaces has increased, especially in urban areas, which destroys wetlands and riparian vegetation and increases runoff. The result is more extreme events, namely greater maximum flows and flooding following high rainfall and lower flows during dry periods (see fig. 6.5).

Dams and dikes affect both the quantity and timing of water flow to

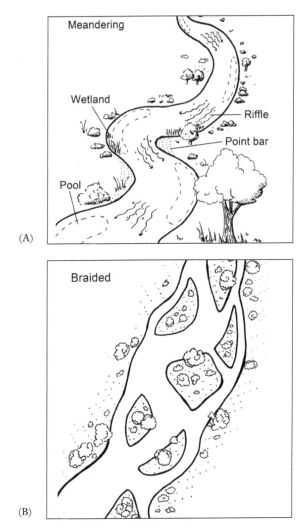

(A)

(B)

Figure 6.6. Different natural channel patterns. (A) Meandering river channel showing different habitat features. (B) Braided river with multiple channels. Drawings by M. Pastor.

many rivers and wetlands worldwide, reducing peak flows that are important for sediment movement and vegetation establishment. As discussed above, dams trap sediments, reducing deposition to the floodplain and associated wetlands. Moreover, dams impede the movement of aquatic organisms.

In addition to changes in flow magnitudes and timing, many streams have been *channelized* to straighten them for navigation and to create land for agriculture or human developments in the areas that were once *floodplains*. In other cases, channel meandering has been restricted by installing *levees* on their banks to reduce flooding. For example, the Kissimmee River in Florida was channelized in the 1960s from a 166-kilometer-long meandering river to a 90-kilometer-long channel to convert two-thirds of the adjacent wetlands to agricultural lands (Kissimmee River case study). Because channelization decreases the heterogeneity of riffles and pools, it increases flow rates, destroys riparian wetlands, and reduces habitat complexity. Channelized rivers are more likely to dry up during low flow periods because the bottoms of channelized rivers are flat and unshaded, thus increasing water temperature and evaporation.

Restoring Hydrologic Regimes in Aquatic Systems

The importance of restoring hydrologic regimes for rivers, wetlands, and lakes cannot be overstated. Many examples show that restoring hydrologic and physical processes can result in substantial recovery because hydrologic inputs are the primary driver of the sediment budget, vegetation dynamics, and habitat quality for flora and fauna (National Research Council 1992; Palmer, Hondula, and Koch 2014). The methods used to restore hydrological processes depend on the size of the river, funding, and to a large degree, social constraints. Because many people live near rivers and we depend on them for water supply, transportation, power, and other uses, restoring river flows nearly always involves lengthy *stakeholder* discussions and balancing different ecological and social values (Elwha River, Kissimmee River, and Sacramento River case studies). Because it is rarely possible to compare different large-scale restoration approaches experimentally, planning river restoration, particularly at the watershed scale, often involves using a variety of models to predict the effects of different restoration options on water flow rates, channel patterns, and sediment budgets (Roni and Beechie 2012). For example, several water flow models were compared in the Sacramento River case study to help select the distance to set back the levee from the river to restore riparian habitat and, at the same time, reduce flooding risk to a town on the river's edge. In addition, eco-hydrological models were used to evaluate and optimize the positive effects of water flows over multiple years on several species of fish, banks swallows, and cottonwood *recruitment* (Alexander et al. 2018). Likewise, a few different modeling approaches were used to predict sediment movement into different habitats following dam removals in the Elwha River case study.

Reducing water withdrawals is an obvious approach to restore the hydrologic regime in aquatic ecosystems. In the past few decades, numerous methods have been developed in countries worldwide to quantify minimum environmental flows necessary to maintain an acceptable level of the desired species and ecosystem functions (Tharme 2003), but the degree to which they are legally enforced varies a great deal across countries (Brierley and Fryirs 2008). One successful example is the recovery of dune slacks, a type of seasonally flooded wetland in the Netherlands, by stopping groundwater pumping and increasing the water inputs on which these wetlands depend (Grootjans et al. 2002). In eastern California, large quantities of water were diverted from natural ecosystems to the city of Los Angeles in the first half of the twentieth century; lengthy legal battles over the past several decades have led to the restoration of minimum flows to the Owens River and to the tributaries to Mono Lake and the lake itself, all of which provide important migratory bird overwintering habitat in these arid ecosystems (Mazaika 2004; Inyo County Water Department n.d.). Successful efforts to halt or substantially reduce water withdrawals are few, however, given increasing human demand for water and changing precipitation patterns.

A second approach to restoring natural flow rates is the removal of dams or other barriers to water flow, particularly in places where sediment buildup behind a dam has reduced water storage and electricity production (see dam removal videos in online resources). Two of the largest dams removed were the Elwha Dam (64 m tall) and the Glines Canyon Dam (33 m tall) on the Elwha River in Olympic National Park in Washington (Elwha River case study). These dams were removed in part to restore the movement of salmon upstream (*Oncorhynchus* spp.), an important resource for the local Native American tribe (Gelfenbaum et al. 2015). In this case, dam removal resulted in the restoration of river channel patterns and streambed substrate, as well as an increase in wetland habitat due to sediment deposition along the coast. Likewise, the Bolsa Chica wetlands in southern California were restored by first excavating and moving sediments, removing oil drilling infrastructure, and then creating a bridge along a coastal highway to hydrologically reconnect the wetlands to the ocean (Southern California Wetlands Recovery Project 2018); the restoration of tidal flow has led to the recovery of wetland plant communities and a diversity of fish and bird species. Numerous small-scale projects worldwide also aim to improve hydrologic connectivity by removing small dams along streams or installing or enlarging *culverts* where streams intersect roads.

When it is impossible to remove dams or other barriers due to social or political constraints, periodic releases of water may simulate high flow conditions and temporarily restore hydrological processes. This approach has been used on the Colorado River, the Murrumbidgee River in Australia, and the Tarim River in China (Glenn et al. 2017). Even short-term water releases that are below natural peak flows can have significant benefits, such as improving riparian vegetation recruitment and increasing sediment inputs below the dams. The effects of water releases are short-lived, however, and the recruitment and survival of riparian trees is low without the restoration of a natural hydrological regime (Docker and Robinson 2014). *Fish ladders* (a series of pools built like steps; see photos in online resources) or bypass channels can be used to allow fish an alternative route around dams, if that is the primary restoration goal. These approaches are less ecologically desirable than complete removal of dams, however.

Whereas many restoration efforts focus on increasing water flow, restoration efforts in areas with compacted soils and paved surfaces often aim to reduce peak flow and associated erosion (Nilsson et al. 2018). These efforts can include restoring heterogeneous topography and increasing infiltration into soil (discussed earlier in this chapter). Alternatively, retention ponds or small wetlands can be installed below mining or construction sites to reduce flows and capture sediments. In urban areas, natural infrastructures, such as vegetated *swales* or *buffer strips*, are used to capture and slow water during high flow events, resulting in the deposition of sediments and associated nutrients and pollutants. The same logic supports the use of in-stream structures, such as *check dams* (also referred to as weirs), to slow water flow, increase water depth, and prevent channel downcutting and bank erosion along streams (fig. 6.7A). Restoring riparian and wetland vegetation helps reduce peak rates, increase sediment deposition, and improve water quality. *Reforestation* with both native and nonnative woody species, however, often reduces the overall water yield in dry periods of the year, particularly in the first few years of a project, because rapidly growing young trees transpire a large amount of water (Filoso et al. 2017).

Restoring Channel Patterns and Riverine Habitat

The best approach to restoring channelized rivers is to remove levees and other structures that block connectivity between the river channel and its floodplain and then allow natural processes to occur. As the channel returns to a natural flow path over time, the dynamic habitat mosaic that is typical of rivers (e.g., pools, riffles, point bars, and riparian wetlands; see fig. 6.6) is re-created (Roni and Beechie 2012). For example, along some stretches of the upper Sacramento River, nonprofit organizations have purchased land and allowed levees to erode, which has resulted in bank erosion, channel movement, and point bar establishment over time, all of which are natural river processes (Sacramento River case study).

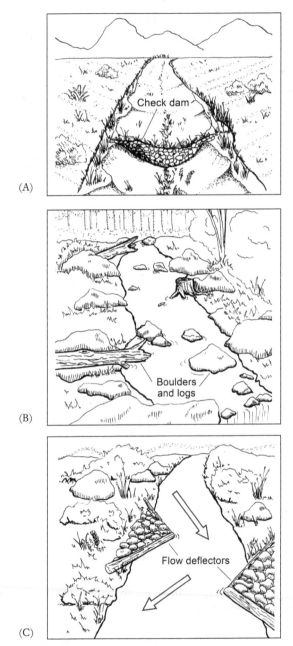

Figure 6.7. In-stream restoration structures. (A) Check dam to reduce flow and channel downcutting in a seasonal stream. (B) Boulders and logs placed in rivers to slow flow and increase habitat heterogeneity. (C) Flow deflectors to increase the *sinuosity* of a stream. Drawings by M. Pastor.

Actively recontouring channelized rivers to increase the *sinuosity* of the river is typically a costly and complicated undertaking but results in many benefits, including restoring habitat, reducing flooding, and improving water quality, because the water flow rate is slower in sinuous rivers and thus permits more filtering through riparian vegetation. For example, it has taken nearly $800 million and fifty years to reconstruct the sinuous channel pattern of the Kissimmee River in Florida, which has restored more than 100 square kilometers of river-floodplain ecosystem (Kissimmee River case study). Likewise, restoring channel sinuosity and floodplain connectivity along stretches of various European rivers has increased habitat complexity and reduced downstream flooding (Kronvang et al. 1998).

To date, most river restoration projects have focused on reconfiguring the channel over short stretches of the river (Palmer, Hondula, and Koch 2014), many following the natural channel design method of Rosgen (1998). Although these methods may improve local habitat conditions over the short term and are often the only viable approach in urban areas (Riley 2016), they do not restore hydrologic processes and sediment loads. Hence, they have been described as "renaturalization" efforts (Wohl, Lane, and Wilcox 2015) that have not fully restored aquatic *biodiversity* (Palmer, Hondula, and Koch 2014).

In smaller rivers and streams, a range of in-stream structures is used to direct water flow (fig. 6.7B, C). Boulders, logs, or flow deflectors can be installed to direct water flow with the goal of narrowing and deepening flow paths, improving bank protection, and creating shelter or breeding grounds for aquatic organisms. Adding gravel is a temporary approach to improve in-stream fish spawning habitat in sediment-starved rivers. Finally, many riparian restoration efforts on smaller streams focus on improving adjacent land management practices, such as fencing out livestock grazing, removing *invasive species*, or revegetating with riparian species known to stabilize banks, reduce local erosion, slow flow, and provide shade along streams. Although these practices can benefit a river or stream on their own, they should be combined with efforts to restore watershed-scale flows and connectivity whenever possible.

Recommended Reading

Craft, Christopher. 2015. *Creating and Restoring Wetlands: From Theory to Practice*. Amsterdam: Elsevier.

>   Thoroughly discusses challenges to and strategies for restoring various wetland types. Chapter 2 provides a short overview of wetland ecology.

Galatowitsch, Susan M., and Joy B. Zedler. 2014. "Wetland restoration." In *Ecology of Freshwater and Estuarine Wetlands*, edited by D. P. Batzer and R. R. Sharitz, 225–60. Berkeley: University of California Press.
Summarizes the process for planning and implementing wetland restoration projects.

Palmer, Margaret A., Kelly L. Hondula, and Benjamin J. Koch. 2014. "Ecological restoration of streams and rivers: Shifting strategies and shifting goals." *Annual Review of Ecology, Evolution, and Systematics* 45:247–69.
Reviews past and current approaches to river restoration.

Roni, Philip, and Timothy Beechie. 2012. *Stream and Watershed Restoration: A Guide to Restoring Riverine Processes and Habitats*. Oxford: Wiley.
Provides a thorough guide to the basic science, planning, implementation, and evaluation of watershed-scale stream restoration projects.

Whisenant, Steven G. 1999. *Repairing Damaged Wildlands: A Process-Oriented, Landscape-Scale Approach*. Cambridge: Cambridge University Press.
Serves as a comprehensive guide to restoring water and nutrient cycling and vegetation in arid and semiarid terrestrial ecosystems by building on natural processes.

7

Soil and Water Quality

Changes in *landform* and *hydrology* affect soil and water quality and nutrient *cycling* both directly and indirectly. When heavy machinery is used to level or excavate land, soil is moved. Inevitably, the soil layers are mixed and altered, affecting *soil texture, soil compaction,* and nutrient availability. As discussed in chapter 6, changes in *topography,* combined with soil compaction, often lead to altered runoff patterns, which in turn can cause further erosion and the movement of nutrients into nearby water bodies, disrupting nutrient *cycling* in both terrestrial and aquatic systems. In this chapter, I provide background and discuss approaches for restoring soil and water quality. I close by summarizing strategies to reduce acidity and toxicants in both terrestrial and aquatic systems.

Humans have made staggering direct changes to global and local chemical cycles (Schlesinger and Bernhardt 2013). The quantity of biologically available nitrogen has more than doubled in the last century, and phosphorus inputs have increased substantially due to the widespread use of fertilizers, emissions from fossil fuel combustion, sewage effluents, and other causes. Carbon dioxide concentrations in the atmosphere have increased dramatically in the past few decades due to fossil fuel emissions and land use conversion, altering the global carbon cycle. In addition, humans have increased the concentrations of many toxic chemicals, such as heavy metals, dioxins, and radioactive materials, to levels that are harmful to themselves as well as other organisms in places across the globe. Certain human activities, such as mining, can expose different rock layers, increasing soil and water acidity and mobilizing a variety of toxic substances. Furthermore, nitrous oxide emissions from power plants and metal smelting operations have increased soil and water acidity in some regions.

Two overarching principles underlie successful efforts to restore soil and water quality. The first priority is to reduce chemical and nutrient inputs before treating the consequences. Second, restoration of water quality in aquatic systems requires the appropriate *management* and restoration of adjacent terrestrial systems. As long as elevated chemical inputs to either terrestrial or aquatic systems continue, any effort to restore the *ecosystem* will require ongoing intensive management to ameliorate these inputs, and restoration will rarely be possible.

Soil Texture, Chemistry, and Biology

Soils consist of four major components: mineral materials, *organic matter*, water, and air. Most undisturbed terrestrial systems have developed a series of soil layers due to the physical, chemical, and biological *weathering* of rock and the deposition of sediment and organic matter. The depth and texture of soil layers vary depending on the geological and climatic history of the site, but the upper layers are typically more fertile and contain more organic matter than deeper layers (fig. 7.1). Soil organic matter is important in nutrient cycling, holding water, soil aeration, and binding the soil together to resist erosion; it also holds a substantial pool of stored carbon (Marin-Spiotta and Ostertag 2016). When human activities disturb soil layers, a number of negative effects result: soil layers are homogenized, water flow patterns change, and organic matter is reduced.

Soil particles are classified into three categories by their diameter size: sand (greater than 0.02 millimeter), silt (0.02 to 0.002 millimeter), and clay (less than 0.002 millimeter). The relative percentage of each of the three particle sizes varies across soil types and is referred to as the *soil texture*. Soil texture strongly affects the movement of water and air through soil, nutrient availability, and how well the soil binds together. For instance, soils with more sand generally drain more quickly than other soils. Higher percentages of clay and silt are associated with greater soil fertility because smaller clay particles bind nutrients. There is no "correct" soil texture, but each restoration site will have a historical soil type, as well as vegetation and soil fauna that are adapted to that soil type.

Plants depend on a host of nutrients, including nitrogen (N), phosphorus, potassium, calcium, magnesium, and sulfur. Nitrogen and phosphorus most commonly limit plant growth. N_2 gas represents 78 percent of Earth's atmosphere, but this form of nitrogen cannot be used directly by

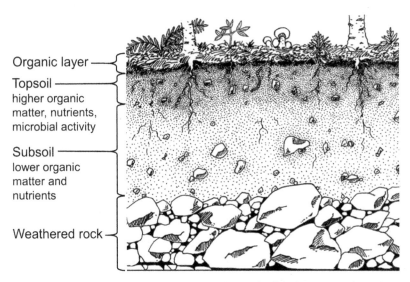

Figure 7.1. Soil layers in an intact ecosystem. The depth of the layers varies by soil type, and each general layer is often divided into sublayers.

plants. Most of the nitrogen used by plants and animals is in the form of ammonium (NH_4^+) or nitrate (NO_3^-). Transformation from nitrogen gas into ammonium is an energy-intensive process that is performed by a few types of bacteria that are either free-living or form *mutualisms* with certain plant species (fig. 7.2); *nitrogen-fixing plant species*, commonly in the bean family, provide energy to these bacteria in return for nitrogen. Phosphorus and most other plant nutrients become biologically available through weathering of rocks. As plants die and their litter decomposes, nitrogen, phosphorus, and other nutrients become available to plants. Soil acidity (discussed below) strongly affects nutrient availability.

Soils host a diverse *biota*, including larger soil fauna (e.g., earthworms, ants, fly larvae, millipedes, burrowing rodents), as well as microscopic organisms such as protozoans, algae, fungi, and bacteria. Together, soil organisms play important roles in decomposition; nutrient cycling; increasing soil porosity, aeration, and water infiltration; binding soil particles together; and in some systems, increasing primary production (Whisenant 1999). For example, *mycorrhizae* are fungi that form mutualistic association with plants. Mycorrhizae increase root surface area and secrete enzymes that help mobilize soil nutrients, particularly phosphorus, as well as enhance water uptake. *Biological soil crusts* are collections of fungi, lichens, cyanobacteria, bryophytes, and algae in varying proportions that form a

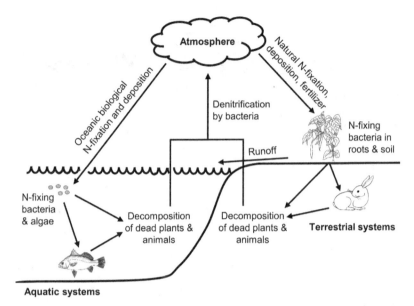

Figure 7.2. Fluxes of nitrogen between the atmosphere, land, and water. Nitrogen fixation is the transformation of nitrogen gas (N_2) from the atmosphere to ammonium (NH_4^+). Ammonium can be transformed into nitrate (NO_3^-); both of these forms are usable by plants and animals. Denitrification is the reverse process of transforming nitrate into nitrogen gas. Clip art licensed under Creative Commons. Figure by J. Lesage and K. Holl.

crust on the soil surface in some arid ecosystems and can fix nitrogen and reduce soil erosion. Disturbing soil layers negatively affects soil organisms and, in turn, the various *functions* they provide.

Restoration of Soil Texture, Nutrient Cycling, and Microbial Communities

The depth and number of soil layers, soil texture, and nutrient availability vary naturally across soil types. Therefore, restoration efforts should consider the intensity and type of anthropogenic *disturbance* to the system and focus on matching the characteristics of the predisturbance soil type to the desired vegetation community (Whisenant 1999). For example, in highly degraded sites like mines, soils are highly compacted and low in nutrients, so it is necessary to increase both organic matter and soil nutrients before establishing vegetation. In contrast, in agricultural landscapes with elevated soil fertility and an abundance of *invasive* plants that are well adapted to take advantage of the high nutrient availability (chap. 8), restoration efforts focus on reducing nutrient levels.

Restoring appropriate soil texture is critical to restoring nutrient cycling processes and therefore to overall restoration success. This point is clearly illustrated by a failed wetland restoration effort in southern California (Boyer and Zedler 1998). The restoration aimed to restore tall cordgrass (*Spartina foliosa*) as *habitat* for an *endangered* bird, the Ridgway rail (*Rallus obsoletus*), but the resulting cordgrass height was much shorter than necessary to serve as rail habitat. After a series of experiments, scientists discovered that the texture of the soil imported to construct the salt marsh was too coarse and infertile for the cordgrass to grow tall and adding fertilizer only increased growth immediately following application because the soil lacked the fine particles to bind and retain the nutrients over the long term.

Ideally, restoration should aim to restore soil layers and texture similar to their predisturbance condition to restore both nutrient cycling and water infiltration processes. However, it is impossible to separate soil layers once they have been mixed by disturbance, making it highly challenging to restore sites where the layers have been disturbed. When human impacts can be foreseen, such as mining, the best practice is to remove and *stockpile* the soil in layers. The soils are then replaced in the original order following mining. In other cases, such as some wetland *mitigation* projects, soil is moved directly from a site that is being disturbed to another site that is being restored. Stockpiling or directly moving soils has the added benefits of reintroducing soil microorganisms and seeds (chap. 9). The duration for which soil is stockpiled should be minimized because the viability of seeds and microorganisms begins to decline after a few months. The dramatic effect of restoring soil layers with the correct texture on vegetation recovery can be seen following sand mining in Western Australia. In a site where the soil layers were removed for mining, mixed, and then replaced, a hard surface soil layer developed and minimal vegetation established (fig. 7.3A). In contrast, a diverse coastal shrubland community established when soil layers were stored separately and replaced in the correct order, resulting in higher nutrient availability and better drainage (fig. 7.3B). Although replacing soil layers dramatically enhances recovery, it is done in only a small number of projects; in most cases, soil layers were disturbed long before the restoration project was planned. Moreover, respreading soil is costly.

In heavily disturbed soils, increasing organic matter is important to improve water holding capacity, nutrient retention, soil aeration, and *carbon storage*, as well as to moderate soil temperature fluctuations (Marin-Spiotta and Ostertag 2016). Many of the erosion control strategies listed in chapter

(A)

(B)

Figure 7.3. Kwongan heath restoration sites in Western Australia. Both photos show twelve-year-old sites that were revegetated with a similar species mix following sand mining. (A) A site where the soil layers were mixed and a hardpan (a dense layer of soil that is largely impervious to water) formed at the surface, preventing vegetation establishment. (B) A site where the soil layers were stored separately and replaced in the original sequence, resulting in successful vegetation establishment. Photos by K. Holl.

6, such as straw, erosion control cloth, and wood mulch, also increase soil organic matter. These techniques trap fine soil particles and nutrients moved by both wind and water, thereby increasing nutrient availability. Compost, animal wastes, straw, hay, or sewage sludge are frequently applied to disturbed soils to increase organic matter and nutrient levels (Bradshaw and Chadwick 1980). Sewage sludge should be tested to ensure that it does not have excessive concentrations of toxic substances, such as heavy metals or pharmaceutical residues. Fertilizers are often applied at the time of seeding or planting to enhance vegetation establishment (chap. 9), but repeated fertilizer application is resource intensive, may elevate nutrient runoff, and tends to favor invasive plant species. It is typical to include nitrogen-fixing plants in the planting mix in low-nutrient sites because they increase nitrogen availability when they shed leaf litter and die (chap. 9).

In sites where nitrogen or phosphorus levels are elevated due to agricultural inputs or atmospheric deposition, the first step is to reduce nutrient inputs. Decreasing in-soil nutrient levels is challenging. Reestablishing *native* vegetation in former agricultural lands gradually lowers available nitrogen and phosphorus (Rosenzweig et al. 2016). Small-scale experiments have shown that repeated application of materials with a high carbon-to-nitrogen ratio, such as sugar, sawdust, or wood mulch, can reduce soil nitrogen concentrations, but these approaches are resource intensive and impractical at a large scale (Baer 2016). Other approaches to reduce nitrogen and phosphorus levels that have met with mixed success include carefully timed burning or repeated mowing and removal of biomass in grasslands (Baer 2016).

Restoring nutrient cycling in degraded lands over the long term requires restoring microbial communities, yet little is known about the recovery of soil fauna and microbes. It is often assumed that soil microbial communities will recover and colonize on their own, which happens in some cases (Scott, Baer, and Blair 2017). In contrast, other research suggests that forest and wetland microbial communities and associated nutrient cycling processes do not recover fully even after multiple decades (Moreno-Mateos et al. 2012; Bonner et al. 2019). The most effective way to reintroduce soil microbial communities is to replace soil that was stockpiled from the site prior to disturbance or to introduce small amounts of soil from intact habitats; this approach helps reintroduce microorganisms and soil fauna and can influence the *trajectory* of plant *succession* (Wubs et al. 2016). Likewise, applying slurries of microorganisms made from damaged soil crusts enhances soil crust recovery in arid systems of the southwestern United States (Chiquoine, Abella, and Bowker 2016). Another approach to reintroduce microbes is to place a small amount of native soil or mycorrhizal fungi in plant pots while they are in the nursery (Whisenant 1999). Middleton and Bever (2012) found that adding native prairie soil to nursery plants enhanced the success of later successional species planted to restore prairie in the midwestern United States.

Although reintroducing microbial communities has many potential benefits, not all effects of microorganisms are positive. The many pathogenic species of fungi and bacteria can have detrimental effects on plants and animals. Moreover, even in generally mutualistic relationships, the interaction can range from negative to positive. For example, Allen et al. (2003) found that adding soil from early successional forest to nursery-grown plants enhanced the growth of tree seedlings planted to restore seasonal tropical forest in Mexico. In contrast, adding late-successional soil had variable and in some cases negative effects on seedling growth, which they attribute to the large carbon demand of the later successional mycorrhizal species. Our knowledge of the role of both the positive and negative effects of microbes in ecological restoration is rudimentary and is an important area for future research.

Water Quality

By nature, aquatic systems are open, meaning that most of the sediments, nutrients, and toxicants come into the system from an outside source. These outside sources include *point source* (a single identifiable source of pollutants, such an emitting pipe or factory smokestack) and *nonpoint source* (many diffuse sources, such as runoff from land, dissolved toxins in precipitation, or atmospheric deposition) *pollutants*. These inputs become increasingly problematic moving down the *watershed*, and the effects are most acute in the lakes and estuaries where they accumulate.

A particular problem in lakes, as well as some slow-moving rivers and nearshore marine ecosystems, is *eutrophication*. Eutrophication is an increase in the supply of organic matter to an ecosystem, often caused by the nutrient enrichment of waters, most commonly phosphorus, beyond their natural levels. Because phosphorus is commonly a limiting nutrient in aquatic systems, elevated phosphorus levels quickly increase the biomass of aquatic plants and cyanobacteria (Gulati, Pires, and van Donk 2012). As the plants die and decompose, the oxygen level declines, which in extreme cases can lead to the death of aquatic organisms. Eutrophication has both negative ecological and economic consequences. Pretty et al. (2003) estimated the cost of eutrophication in England and Wales to be $105 million to $160 million per year due to reduced property values, drinking water treatment, and loss of recreational value.

A related problem is increased sediment inputs to water bodies. *Channelization* of rivers and the destruction of wetland and *riparian* vegetation increase the amount and size of sediments transported by rivers (chap. 6). Suspended sediments decrease water clarity, which reduces plant photosynthesis and harms many filter feeders. When sediments settle out in rivers, they alter streambed texture, which affects the eggs and larvae of fish and other aquatic species. Sediment deposition in lakes reduces lake depth and water volume. In addition, nutrients, organic matter, and some pollutants bind to fine sediments and are transported through the watershed with them, thereby exacerbating eutrophication when the sediments settle out in aquatic systems. Heavy metals, such as copper, zinc, and lead, can accumulate to toxic levels in lakes and wetlands over time when they are washed in with increased sediment loads.

Restoration of Water Quality

Successful restoration of water quality depends almost entirely on (1) reducing both point and nonpoint source inputs and (2) improving land-management practices and restoring terrestrial and wetland habitat throughout the watershed to retain nutrients and sediments. Many countries have policies to regulate nutrient inputs from agriculture, sewage, and fossil fuel combustion (chap. 11; McCrackin et al. 2017). In some cases, the enforcement of such laws has led to dramatic decreases in nutrient levels and increases in water clarity. For example, Lake Washington near Seattle was highly eutrophied in the 1960s due to effluent inputs from several waste treatment plants. After a successful public referendum, 99 percent of the waste was diverted or treated, which by the early 1970s led to the recovery of water clarity and a dramatic reduction in algal blooms (National Research Council 1992). In contrast, a recent review of lake and coastal marine ecosystem recovery following nutrient reduction found that the majority of the eighty-nine case studies had not fully recovered to baseline levels, suggesting that recovery from eutrophication is a slow process and does not happen in all cases (McCrackin et al. 2017).

Nonpoint source pollutants are much more difficult to control than point sources because their causes are more diffuse. As with point source pollution, the first step is to reduce the source emissions through a variety of strategies, such as improving agricultural practices, reducing tailpipe emissions, installing temporary erosion control measures during construction, using low-phosphorus detergents, or educating people about appropriate waste disposal practices. For example, reducing the amount of and carefully timing fertilizer inputs to agriculture, as well as leaving crop residues on fields, can reduce agricultural runoff and therefore nutrient inputs into nearby water bodies. In a review of sixty studies in northwestern Europe, Van Vooren et al. (2017) found that *hedgerows* and grass *buffer strips* in agricultural fields reduced nitrogen and phosphorus runoff by two-thirds and trapped 90 percent of sediments in surface flows. In many areas of the United States, extensive educational campaigns have worked to increase awareness that lawn and garden care products, motor oil, pet waste, and household chemicals, if not properly disposed of, all run off into storm drains and bodies of water downstream.

Restoring wetlands and better managing riparian forests reduce nutrient and sediment inputs to nearby water bodies (Roni and Beechie 2012; Baer 2016; Hansen et al. 2018). Both wetlands and riparian forests slow water flow (chap. 6), which results in the deposition of sediments and nutrients that are bound to sediments before they enter nearby water bodies. Moreover, the slower water flow allows more time for plants to take up nutrients and for microbial nutrient cycling to occur (Baer 2016). Wetlands host a range of beneficial microbial processes, such as immobilizing heavy metals and supporting bacteria that reduce nitrate concentrations (Galatowitsch and Zedler 2014). Although wetlands can take up large quantities of nutrients due to their high *productivity* and microbial processes, they do not have an unlimited nutrient uptake capacity. Hence, wetland restoration should not be viewed as a substitute for reducing nutrient and sediment inputs throughout the watershed.

Retention ponds are a method to capture peak stormwater runoff, allowing sediments to settle before slowly releasing the water. *Best management practices* for riparian zones include *revegetation*, installing mulch or erosion control cloth after soil disturbance until vegetation can establish, and prohibiting grazing by livestock near the stream edge (Roni and Beechie 2012). Removing and restoring old logging roads and improving the intersections between roads and streams (e.g., installing *culverts* and bridges) also reduce sediment inputs to nearby streams (Roni and Beechie 2012).

Several techniques are used to manage nutrient and sediment inputs that accumulate in lakes (National Research Council 1992; Gulati, Pires, and van Donk 2012; Cooke et al. 2016). For example, dredging removes excess sediments, but it also causes extensive damage to sediment-dwelling organisms. The salts of calcium, iron, or aluminum are added to some lakes and reservoirs to precipitate excess phosphorus into a biologically unavailable form. In addition, some highly managed lakes and reservoirs may be artificially aerated to increase oxygen, aiding in the uptake of nutrients by microbes. Clearly, however, these methods are expensive, short-term management actions that cause additional disturbance to the system, do not result in a self-maintaining system over time, and do not address the root cause of the problem, namely excessive inputs from outside the lake.

Acidity

Certain human activities increase the acidity (i.e., lower the pH) of soil and water. In mined areas, the exposure and weathering of certain rocks increases soil acidity, and these rocks leach acid mine drainage into nearby water bodies. These effects are most dramatic in areas with heavy metal mining (e.g., copper, zinc, nickel), such as the Sudbury region of Ontario, Canada (Gunn 1995). In addition to acid mine drainage, smelters that process the metals historically emitted large quantities of sulfur dioxide, leading to acid rain; the combination of acid mine drainage and acid rain killed vegetation across thousands of hectares and caused severe damage to lakes up to 30 kilometers from smelters in this and other heavy metal mining regions. Human emissions of sulfur and nitrogen compounds from electricity generation (primarily from coal), factories, and motor vehicles have resulted in acid rain over hundreds of kilometers, affecting both soil and nearby water bodies. These emissions have been particularly problematic for lakes, which collect water from large areas. Fortunately, improved power plant technologies have reduced harmful emissions in many regions in the past few decades. On the other hand, continued extensive anthropogenic carbon dioxide emissions are slowly acidifying seawater across the globe.

Increasing acidity not only negatively affects aquatic organisms directly by killing sensitive aquatic fauna and dissolving high *biodiversity* coral reefs, but the increased concentration of hydrogen (H^+) ions also affects the mobility of other positively charged ions (e.g., potassium, calcium) and various microbial processes (Baer 2016). Hence, higher soil acidity leads to reduced nutrient availability and the increased leaching of positively charged metals (e.g., lead, aluminum, iron), leading to potentially toxic concentrations.

Reducing Acidity

The primary way to decrease acidity is to reduce the inputs that caused the acidification. Restoring mine sites to minimize ongoing acid mine drainage runoff, reducing fossil fuel consumption (particularly coal), and using the best available technologies to limit emissions from power plants are three keys to reducing concentrations of acidifying compounds. Lime ($CaCO_3$), which temporarily reduces acidity, is often applied to acidified soils along with fertilizer to facilitate plant establishment (Bradshaw and Chadwick 1980; Gunn 1995). Lime is also applied in lakes and has been used widely in Sweden and Norway as a temporary solution to reduce lake acidity, but it must be applied on an ongoing basis if emissions are not reduced (Gulati, Pires, and van Donk 2012). Heavily acidified sites are usually revegetated using acid-tolerant plants, a method that increases organic matter and decreases erosion, thereby lowering the acidity of soil and nearby water bodies over time.

Toxic Chemicals

Areas disturbed by mining and industrial activities commonly are contaminated with a variety of toxic chemicals, including those used in dry cleaning, radioactive wastes, pesticides, heavy metals, and oil. A first step in restoring such areas is to clean up the toxic wastes, a difficult and expensive process; billions of dollars are spent globally each year on this task (Pilon-Smits and Freeman 2006). Cleanup can be complicated because multiple toxic substances are often present at a single site, and rarely are the identities of the chemicals present well documented.

Cleaning Up Toxics

Although a detailed discussion of toxic waste cleanup is well beyond the scope of this book, a brief summary of a few of the general approaches can be noted (Pilon-Smits and Freeman 2006). The methods used depend on the specific chemicals present. In extreme cases, highly contaminated soils can be capped in place or removed and stored in a permanent containment structure. In other cases, toxic chemicals are treated on-site. Some chemicals can be removed by washing or venting soil or by extracting the chemical using a solvent. Other compounds can be transformed to less harmful compounds through chemical reactions. *Bioremediation* is the employment of microorganisms or plants to remove, degrade, or immobilize toxic chemicals in soil, water, or air through a range of mechanisms (Pilon-Smits and Freeman 2006). Planting species that are tolerant of the toxic chemicals at a site can stabilize soils and minimize the leaching of pollutants. Some terrestrial and aquatic plants concentrate certain heavy metals. The toxin-loaded plant material can be harvested and removed, which is cheaper and less damaging than removing contaminated soil. Furthermore, plants and microbes can facilitate the breakdown of certain carbon-based compounds. For example, oil spills can be cleaned up by either actively introducing microorganisms or adding nutrients or oxygen to enhance microbial growth and thereby increase the mineralization rate of the oil (Leahy and Colwell 1990).

Toxic waste cleanup is extremely expensive, particularly when there are multiple chemicals at a site. If the chemicals are removed through extraction or biomass harvesting, the question of where to dispose of the chemicals is raised. Toxic waste cleanup is just the first stage of restoration; *abiotic* and biotic conditions will need to be restored following cleanup.

Recommended Reading

Baer, Sara G. 2016. "Nutrient dynamics as determinants and outcomes of restoration." In *Foundations of Restoration Ecology*, 2nd ed., edited by Margaret A. Palmer, Joy B. Zedler, and Donald A. Falk, 333–64. Washington, DC: Island Press.
  Provides a thorough review of the ecology and restoration of nutrient cycling in terrestrial systems.

Gulati, Rumati D., L. Miguel D. Pires, and Ellen van Donk. 2012. "Restoration of freshwater lakes." In *Restoration Ecology*, edited by Jelte Van Andel and James Aronson, 233–47. Malden, MA: Blackwell.
  Reviews different methods to restore water quality in lakes.

National Research Council. 1992. *Restoration of Aquatic Ecosystems*. Washington, DC: National Academy Press.
  Reviews restoration of wetland, river, and lake ecosystems and describes many case studies.

Pilon-Smits, E. A., and J. L. Freeman. 2006. "Environmental cleanup using plants: Biotechnological advances and ecological considerations." *Frontiers in Ecology and the Environment* 4:203–10.
  Provides a succinct overview of different toxic waste cleanup methods.

8

Invasive Species

For millennia, the spread of plant and animal species was limited by their dispersal abilities because humans spent most of their lives in the same place. Recently, human global travel has radically altered species distribution patterns through both accidental and intentional movement of species to areas they did not historically inhabit. Various terms are used for species that are found outside of their historical range, including *nonnative*, exotic, and alien. *Ecological restoration* typically aims to restore *native* species, or species that have evolved in a specific location, but the question of how long a species needs to have inhabited or evolved in a given location to be considered native is subjective. Is a species nonnative if it was introduced by humans within the last one hundred years, two hundred years, or five hundred years? The earlier a species was moved, the less likely there are records of the exact species distribution prior to its transport.

It is important to clarify the distinction between nonnative species and *invasive species*. Invasive species are those that dominate native ecosystems following introduction, cause harm to native species, and alter ecosystem processes. Invasive species are a major *biotic* barrier to ecosystem *recovery*, so their control is a focus of many restoration efforts. Although many species have been moved to new locations by humans and thus are considered nonnative, only a small proportion of them spread into and strongly negatively affect the ecosystems they colonize. Only this subset of introduced, nonnative species is considered invasive.

In addition, there is increasing recognition of native "invaders" (Carey et al. 2012), or native species whose populations have grown substantially within their native range due to human activities and that have detrimental impacts on other native species and ecosystem processes. Native invaders can be species that are well adapted to human *disturbance*, such as the western gray kangaroo (*Macropus fuliginosus*), which has benefited from predator control in Australia, or the various species of cattails (*Typha* spp.) that dominate disturbed wetlands in numerous regions worldwide, reducing native diversity and altering nutrient *cycling* (Carey et al. 2012). In other cases, native invaders are the result of intentional human activities to increase their population sizes, such as stocking lakes and reservoirs with many recreational-fishing species. When native invader species are sufficiently abundant to be a challenge to restoring a native ecosystem, they may be the focus of control efforts.

I start this chapter by discussing how invasive species spread and why they are a problem for restoration. I then review approaches for *eradicating* (removing entirely) or controlling different types of invasive species, both native and nonnative, as part of restoration efforts. I close by briefly discussing controversies regarding invasive species control. I use the term *invasive species* to refer to the impact of the species on the ecosystem regardless of their origin (native vs. nonnative) and use the modifiers native and nonnative when appropriate.

How Invasive Species Spread

Humans have moved plant and animal species from one region to another for many reasons. People have transported species they value to new places intentionally, as in the cases of crop and landscaping plants, trees for forestry, livestock, and pets. In other cases, nonnative plants were introduced for erosion control, such as the case of European beach grass (*Ammophila arenaria*), which was planted extensively along coastal dunes in western North America to stabilize moving sands.

Other invasive nonnative species were introduced accidentally. For example, several common tree and shrub diseases have spread into natural habitats through the nursery and timber trades. Likewise, boats moving through waterways (e.g., the Panama Canal, which connects the Atlantic and the Pacific Oceans) have transported many invasive aquatic species, either attached to the boats or carried in *ballast water* that boats take up or discharge when they unload or load. One reason why invasive nonnative populations increase rapidly in new habitats is that there are fewer predators or pathogens that have evolved to control their populations.

Traits of Invasive Species and Invasible Ecosystems

A number of traits are common to invasive species, although not all invasive species have all these traits. Typically, invasive species have good dispersal abilities or spread vegetatively. Invasive aquatic species, particularly those that are highly mobile in either the adult or larval forms, can spread quickly with the movement of water and boats. Invasive species are often tolerant of a wide range of habitat conditions and grow and reproduce quickly, particularly when light, water, nutrients, and food sources are abundant.

Species that become invasive in a given habitat usually come from a similar ecosystem elsewhere on the globe, which can help identify likely invaders. For example, many common invasive species in coastal California come from other Mediterranean climate regions (e.g., Australia, South Africa, and the Mediterranean basin) and vice versa: the grassland herb California poppy (*Eschscholzia californica*) is invasive in other Mediterranean regions.

To manage an ecosystem for reduced invasibility, it is also important to recognize characteristics of ecosystems that make them susceptible to invasion. Heavily fragmented and disturbed habitats are more susceptible to invasive species because most invasive species are better adapted than native species to use available resources rapidly. Thus, native ecosystems embedded within agricultural landscapes with high nutrient and water inputs are at greater risk of invasion than those in landscapes with minimal anthropogenic disturbance. Likewise, the high light conditions at the edge of forest patches and along trails, combined with increased seed dispersal from peoples' shoes and vehicles, facilitate the colonization and establishment of invasive species. For example, a major obstacle to restoring degraded forest remnants in the Atlantic forest in Brazil is invasive native vines that proliferate under the high light conditions at forest edges and in canopy gaps, carpeting the remaining trees and reducing tree growth and seedling *recruitment* (César et al. 2016).

Invasive nonnative terrestrial species tend to be particularly problematic on islands, where animals and plants have been isolated from predators or *competitors* for millions of years and have not evolved defense mechanisms. For example, many island-dwelling birds have evolved to become flightless and have lost their fear of predators, making them particularly vulnerable to invasive nonnative predators like snakes and mammals. Similarly, island plants often lack defense mechanisms such as spines or chemicals to deter herbivory, so they are susceptible to nonnative grazers like goats and rabbits (Galapagos Tortoise case study).

Why Invasive Species Cause Problems for Native Species and Ecosystems

Invasive plant, animals, and pathogens negatively affect native species and ecosystem recovery in many ways (table 8.1). In some cases, invasive species outcompete native species directly and stop the natural process of *succession*. Invasive and feral predators, such as cats, foxes, snakes, and large lake fishes, also destroy native animal populations through direct predation. Invasive diseases have led to the entire or near *extinction* of several native species of ecological and economic importance. Sometimes, invasive species hybridize with native ones, resulting in the loss of locally adapted *ecotypes* (genetically distinct subpopulations of a species). Finally, invasive species frequently change the *abiotic* conditions or patterns of disturbance in native habitats. These new abiotic conditions often favor additional invaders (D'Antonio, August-Schmidt, and Fernandez-Going 2016).

In many cases, invasive species negatively affect the native habitat through multiple of these mechanisms. For instance, *nitrogen-fixing* acacia trees (*Acacia* spp.) from Australia have spread extensively in other Mediterranean climate regions, such as Chile, Spain, California, and South Africa; these trees increase fire severity due to their high biomass, alter soil microbial communities, reduce water availability as a result of high transpiration rates, and outcompete native plant species (Le Maitre et al. 2011). As another example, smooth cordgrass (*Spartina alterniflora*), native to the eastern United States, was introduced as part of an experimental restoration project in San Francisco Bay. Smooth cordgrass spread widely and hybridized with California cordgrass (*Spartina foliosa*). The hybrid tolerates a range of habitat conditions and has had multiple negative effects, including outcompeting several rare plant species, reducing the area of mudflats that shorebirds use for nesting and feeding, and altering the *hydrology* by clogging drainage channels (Kerr et al. 2016). Invasive lionfish (*Pterois* spp.) destroy populations of native reef fishes, some of which eat algae; loss of reef fishes can lead to overgrowth of algae, which hurts coral reefs.

Invasive species are of concern to more than just *natural resource managers* because of their huge economic impacts. Zebra mussels (*Dreissena polymorpha*), native to Russia and Ukraine and widely introduced in North American water bodies, regularly clog cooling tower intake systems to power plants. One estimate suggests that $500 million is spent annually in removing these mussels and preventing their spread (Hoddle n.d.). Invasive insects, pathogens, and plants can also cause enormous financial losses to agricultural production (Pimentel, Zuniga, and Morrison 2005).

Table 8.1. Examples of Negative Effects of Invasive Species

Species	Location	Negative effect	Citation
Wild sugarcane (*Saccharum spontaneum*)	Latin America	Outcompetes native species and impedes succession on abandoned deforested lands.	Hammond 1999
Black and brown rats (*Rattus* spp.)	Galapagos Islands, Ecuador	Preys on juvenile giant tortoises (*Chelonides nigra*).	Galapagos Tortoise case study
Nile perch (*Lates niloticus*)	Lake Victoria, East Africa	Preys on native fish species.	Ogutu-Ohwayo 1990
Water molds (*Phytopthora* spp.)	Australia, New Zealand, Europe, North America	Diseases that have caused dramatic declines of many plant species and faunal species that eat those plants. Causes economic losses to timber industry, and increases fire management costs due to higher fuel load.	Sims et al. 2019
Grass carp (*Cteno-pharyngodon idella*)	Freshwater bodies in the United States	Alters the abiotic environment in lakes and rivers by stirring up sediments that reduce water quality and the reproductive success of other fish species.	Pípalová 2006
Saltcedar (*Tamarix* spp.)	Southwestern US creeks and rivers	Alters the abiotic environment, reducing groundwater level and increasing salinity, which inhibit the recovery of native vegetation.	Tamarisk Removal case study
Brown trout (*Salmo trutta*)	Rivers in the Adriatic basin of Europe	Outcompetes and hybridizes with the native marbled trout (*Salmo marmarotus*).	Crivelli 1995

Stages of Invasive Species Invasion and Control

The most effective way to control invasive nonnative species is to prevent their spread beyond their native range (fig. 8.1). Doing so means enforcing strict legislation regarding the transport of any potential or known invasive species. Several global, regional, and country-specific efforts have enacted legislation to restrict large shipping vessels from exchanging ballast water near shore to reduce the likelihood of invasive aquatic larvae settling in coastal bays (Firestone and Corbett 2005). New Zealand has some of the strictest laws to prevent invasions (Boonstra 2010), which lead to a high

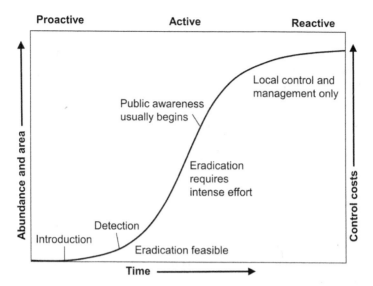

Figure 8.1. Stages of invasive species spread and control. Figure revised from Hobbs and Humphries 1995.

level of coordination across agencies to prevent and rapidly detect invasions. Most countries have a "dirty list" of known invaders that are forbidden for entry, but because this list includes species that are already well established in the country, it is too late to prevent invasion (see fig. 8.1). In contrast, New Zealand has a "clean list" of all species found in the country in 1998, when legislation was passed. Any species not on that list must be screened by the government Environmental Risk Management on Authority to assess its potential risk prior to allowing entry (Boonstra 2010).

Preventing nonnative invasive species spread requires a coordinated early detection and control program. For example, the Early Detection and Distribution Mapping System is an online invasive plant mapping tool used by forty U.S. states and four Canadian provinces (Center for Invasive Species and Ecosystem Health n.d.). This resource serves to alert land managers of invasive plants in their area and help them prioritize removal efforts of those species that are still at an early stage of invasion, when eradication may still be possible.

Education and outreach are important tools to control the spread of invasive species. Many people are not aware of the identities or negative effects of invasive species and often spread them accidentally. For instance, invasive plant seeds are spread on the boots and wheels of hikers and cyclists, some aquatic invasive species have been introduced through dumping aquaria species into sewage systems or waterways, and tree pathogens are often spread through the transport of infested firewood. Public education and restrictions on sales and transport of invasive species can aid in reducing their spread.

Following *best management practices* for controlling pathogens during restoration activities also helps limit their spread. Sims et al. (2019) found that plant nursery sanitation practices were highly effective in eliminating infection by *Phytophthora* spp., a genus of plant pathogens that has had a devastating effect on native species in several Mediterranean regions (see table 8.1); these practices include sterilizing potting soil, not allowing for accumulation of standing water, and cleaning all shoes, vehicles, and tools entering the nursery.

As an invasive nonnative species increases in abundance, eradication becomes more resource intensive (see fig. 8.1). Given the resources and long-term commitment needed to control most invasive species, it is necessary to prioritize invasive species for control efforts (D'Antonio, August-Schmidt, and Fernandez-Going 2016). Criteria that should be considered include:

1. How feasible is control and eradication? Is the species extremely widespread, apt to persist through long-lived seed banks, or likely to generate a public outcry if removed?
2. How severe are the ecological and economic impacts of the species? Typically, more resources are available for those species that have clear negative economic impacts.
3. If the species is removed, is the ecosystem likely to recover without further intervention, or are additional actions needed, such as planting native species?

Once an invasive nonnative species has reached a certain abundance, or if the species is an invasive native species, eradication is not feasible. Then the only option is to keep it at a level that allows for recovery of the native species or processes that are the goal of restoration (see fig. 8.1).

Before beginning to eradicate or manage invasive species, it is important to consider what, if any, additional restoration will be necessary to prevent reinvasion and to ensure that native species and ecosystems recover. Commonly, invasive species reinvade or their populations rebound following control efforts. If eradication efforts target a small area and there are nearby source populations, then the likelihood of reinvasion is high. Moreover, one or a suite of other invasive species may immediately colonize the area in its place (D'Antonio, August-Schmidt, and Fernandez-Going 2016).

In some cases, the invasive species have so substantially transformed the ecosystem that it is unlikely to recover unless the abiotic conditions are also manipulated following invader removal (D'Antonio, August-Schmidt, and Fernandez-Going 2016). For example, control of the invasive tree *Tamarix* spp. in *riparian* areas of the southwestern United States has not led to recovery of native riparian trees in areas where water availability is low, soils have become saline, and nearby native seed sources do not exist (Tamarisk Removal case study). If *propagules* of native species are not on-site and are unlikely to disperse from a nearby population, then invasive removal must be combined with native species *reintroduction*. D'Antonio, August-Schmidt, and Fernandez-Going (2016) suggest introducing native species that have similar traits as invaders, such as fast growth and good competitive abilities, so that they can successfully outcompete the invasive species.

Another consideration before undertaking invasive control efforts is the need for long-term commitment to *monitoring* and *management*. Otherwise, extensive resources may be expended on invasive control with little long-term impact. Van Wilgen et al. (2012) found that despite the South African government spending $457 million between 1995 and 2010 to remove invasive trees, the overall cover of those species increased, due in part to poor monitoring and follow-up on removal efforts. To truly eradicate an invasive species, not only should all individuals be removed at a single time, but any individuals that reinvade or reemerge need to be removed. Some invasive plants have seed banks that last fifty or more years, such as French broom (*Genista monsspesulana*), a common invasive shrub in Mediterranean-type ecosystems. In these cases, removal efforts for a handful of years are insufficient, and long-term maintenance and monitoring will be needed for decades.

Methods for Eradicating and Controlling Invasive Species

Methods used to eradicate and control invasive species vary with the species biology, the size of the invasion, and the availability of resources. It may be feasible to manually remove an invasive plant species over a small area when there is an extensive volunteer base, but this approach is unlikely to be practical for widespread invaders or when resources are limited. The method used for removal also depends on constraints of the site. For instance, the use of controlled burns is restricted near urban areas due to concerns about accidental spread to human infrastructure and effects on air quality. It is also important to consider negative consequences of control methods. Broad-spectrum herbicides may negatively affect nontarget species or raise concerns about water contamination, for example, and removing feral animals by poisoning their food supply can kill nontarget organisms. Most successful invasive control efforts use an *integrated pest management* approach that focuses on long-term control of invasive species or their damage through a well-planned combination of physical removal, biological control, and habitat manipulation when possible. Chemical control methods are used only after careful consideration of other methods. In addition to the most common approaches discussed below, many specialized invasive control techniques have been developed for individual species.

Physical Removal

Various physical removal techniques are used to control invasive plants and animals, and restoration *practitioners* have developed specialized mechanical tools and methods to expedite the work (Holloran et al. 2004). Some invasive plant species are manually removed by cutting or pulling. César et al. (2016) found that cutting invasive native vines in degraded tropical forest remnants in Brazil increased growth of a diversity of native seedlings, small trees, and shrubs, thereby *facilitating* forest recovery. Some invasive animals have been controlled using trapping, hunting, or fishing. Such efforts have been particularly effective on islands where reinvasions are easier to control (Jones et al. 2016). In ponds or lakes, electroshocking has been used to stun and then physically remove invasive fish species when large-scale chemical treatments are too extreme or harmful (Britton, Gozlan, and Copp 2011).

Chemical Control

Numerous herbicides, pesticides, and fungicides are used to control invasive plants, animals, and pathogens. In some cases, chemicals are applied over large areas, such as aerial spraying of invasive cordgrass hybrids across large areas of San Francisco Bay (Kerr et al. 2016). It is preferable, however, to apply chemicals more locally, such as by first cutting invasive wood species and then painting herbicides directly on the stumps to prevent resprouting (Holloran et al. 2004). Poison baits are used to control invasive fauna, such as rats. Although chemical control is often the most cost-effective control method, the pros and cons of this method and alternatives should be evaluated. Some municipalities restrict the use of some or all chemicals. Moreover, species can develop resistance to chemicals after repeated and widespread use. Herbicides, pesticides, and fungicides should be applied in a manner that minimizes risks to human health, nontarget organisms, and soil and water quality.

Biological Control

Biological control agents have been used successfully to control invasive species in some cases. For instance, sterile grass carp (*Ctenopharyngodon idella*) have been introduced to lakes to consume a variety of invasive aquatic plant species and are more cost-effective than physical or chemical removal methods (Natural Resources Council 1992). Saltcedar leaf beetles (*Diorhabda* spp.) have reduced the cover of invasive *Tamarix* spp. in much of the southwestern United States (Tamarisk Removal case study). Biological control agents should be thoroughly tested prior to widespread release to evaluate their effects on nontarget species. In some cases, insect and disease biological control agents have spread beyond the intended species to dramatically reduce populations of closely related native species (Louda and O'Brien 2002). For example, the mongoose (*Herpestes auropunctatus*), a small carnivorous mammal from Asia, was introduced into Hawaii to control rats in Hawaiian sugarcane plantations. However, because the mongoose is active during the day and rats are nocturnal, the mongoose ate native bird eggs and destroyed native bird populations.

Managing the Ecosystem to Favor Native Species

Managing the disturbance regime (chap. 5) can reduce the abundance of invasive species in some ecosystems and shift the balance toward the desired native species. For example, properly managed cattle or elk grazing reduces the cover of tall-statured invasive grasses and thereby improves the success of low-statured native annual wildflowers in California prairies (Stahlheber and D'Antonio 2013). Likewise, restoring a natural flooding regime in wetlands or riparian systems often favors native species (Stromberg 2001). The success of efforts to manage invasive species by altering disturbance regimes depends on the timing of the disturbance, as well as the adaptations of both the invasive and focal native species to disturbance. Because invasive nonnative species are commonly from habitats with similar disturbance regimes, they may be equally or better adapted to a disturbance regime (e.g., fire and many invasive grasses; D'Antonio and Vitousek 1992) that otherwise benefits native species.

Invasive species are often successful because they can take advantage of high light, nutrient, water, and food availability, so some restoration efforts focus on reducing these resources to favor native species that compete better under low resource conditions. The repeated removal of aboveground biomass for a few years prior to restoration may reduce elevated soil nitrogen levels (chap. 7; Baer 2016), although this approach is resource intensive.

Draining artificial water sources, such as cattle ponds, has reduced populations of some invasive predatory frogs and toads, which prey on native salamanders, smaller frogs, and other small animals.

As discussed earlier, combining invasive removal with the introduction of target native species that fill niches occupied by native species reduces reinvasion (D'Antonio, August-Schmidt, and Fernandez-Going 2016). Planting native vegetation to prevent reinvasion has been a key to successful riparian habitat restoration in locations where invasive tamarisk trees have been killed by beetles or have been removed (Tamarisk Removal case study).

Controversies over Invasive Species Management

Despite the clear ecological and economic benefits associated with controlling most invasive species, such restoration efforts can be controversial for many reasons. First, some invasive species are highly valued by humans for cultural and economic reasons. For example, the Asian swamp buffalo (*Bubalus bubalis*), which was introduced to northern Australia in the 1800s, does considerable damage to freshwater ecosystems and carries diseases that are a threat to the livestock industry, but they are a source of food and income, through meat and hide sales, for some Aboriginal communities, making their eradication controversial (Collier et al. 2011). Moreover, animal rights activists often raise concerns about invasive faunal eradication efforts, particularly if methods used are considered inhumane, as in the case of the attempted removal of invasive grey squirrels via trapping and euthanasia in Italy (Perry and Perry 2008).

Some invasive species have positive effects on native ecosystems and species, such as increasing *carbon storage* and facilitating establishment of native species in highly disturbed sites (Norton 2009; Schlaepfer, Sax, and Olden 2011). Tablado et al. (2010) found that the invasive nonnative crayfish (*Procambarus clarkii*) caused declines in native crayfish, amphibians, and invertebrates in marshes in southwestern Spain, but also led to an increase in the abundance of native predatory birds that eat the invasive crayfish. In highly altered ecosystems where native species, such as large seed-dispersing birds, have gone extinct, nonnative species may fill those roles.

These complex effects of invasive nonnative species on native species and ecosystems have led to heated and ongoing debates in both the practitioner and the academic communities. For example, control of invasive *Tamarix* spp. along riparian areas in the southwestern United States has been controversial because *Tamarix* provides habitat for the *endangered* southwestern willow flycatcher (*Empidonax traillii extimus*) (Tamarisk Removal case study; Dudley and Bean 2012). Some scientists (e.g., Davis et al. 2011; Schlaepfer, Sax, and Olden 2011) have argued that once invasive species are well established, both their negative and positive impacts on native *biodiversity, ecosystem services,* and human health should be carefully weighed before allocating resources for control. Others (e.g., Vitule et al. 2012) argue that invasive nonnative species have net negative impacts on native ecosystems and that it is difficult to calculate their risks particularly over the long term. Discussion about how to manage native invaders is equally heated because management decisions necessarily require prioritizing certain native species over others (Carey et al. 2012).

Preventing introductions in the first place is clearly the most effective and least controversial approach to controlling invasive species. If nonnative species are actively introduced for biological control or to fill the role of an extinct native species, then the risks should be weighed carefully and tested at a small scale prior to implementation. Once invasive nonnative species are well established, there are no easy answers about when and how to control them, and often there is disagreement among *stakeholders* about the best approach. In each case, the goals of the project, restoration options and constraints, the extent of the invasion, and ecological, socioeconomic, and cultural factors must be weighed to determine how much effort to invest in trying to control a species and to select one or a suite of methods to do so. Consulting with varied stakeholders during the planning phase and informing them of the rationale for the selected approach may minimize controversies and ensure successful implementation (chap. 3; Crowley, Hinchliffe, and McDonald 2017). It is also important to coordinate control efforts across different natural resource managers in an area to prevent rapid reinvasion from nearby source populations.

Recommended Reading

Carey, Michael P., Beth L. Sanderson, Katie A. Barnas, and Julian D. Olden. 2012. "Native invaders—Challenges for science, management, policy, and society." *Frontiers in Ecology and the Environment* 10:373–81.
    Discusses the issue of native invasive species and challenges for management.

D'Antonio, Carla M., Elizabeth August-Schmidt, and Barbara Fernandez-Going. 2016. "Invasive species and restoration challenges." In *Foundations of Restoration Ecology*, 2nd ed., edited by M. A. Palmer, J. B. Zedler, and D. A. Falk, 216–44. Washington, DC: Island Press.
    Provides an overview of ecological concepts related to invasive species and restoration.

Holloran, Pete, Andrea Mackenzie, Sharon Farrell, and Doug Johnson. 2004. *The Weed Workers' Handbook*. Richmond, CA: California Invasive Plant Council.
Serves as a practical guide to leading volunteer plant invasive removal projects, with thorough descriptions of methods and tools.

Schlaepfer, M. A., D. F. Sax, and J. D. Olden. 2011. "The potential conservation value of nonnative species." *Conservation Biology* 25:428–37.

Vitule, J. R. S., C. A. Freire, D. P. Vazquez, M. A. Nuñez, and D. Simberloff. 2012. "Revisiting the potential conservation value of nonnative species." *Conservation Biology* 26:1153–55.
Schlaepfer, Sax, and Olden 2011 and Vitule et al. 2012 debate the positive and negative impacts of nonnative species and implications for their management.

9

Revegetation

Restoration of terrestrial ecosystems typically focuses on restoring plant communities, which provide important ecosystem services such as nutrient *cycling* and erosion control and are habitat for faunal communities. Likewise, recovery of kelp, seagrass, and various wetland plants is key to the restoration of many aquatic ecosystems. I begin this chapter by discussing whether and when to actively *reintroduce* vegetation as part of restoration efforts. I then discuss considerations for selecting plant species and genetic material, as well as options for how to propagate plants. Finally, I describe methods for enhancing plant survival in restoration projects.

Tailoring the Revegetation Strategy to the Site

An important step in the planning process (chap. 3) is to consider the *goals* of the project, the resources available, and the natural resilience of the ecosystem to determine the most appropriate and effective way to facilitate vegetation *recovery* (fig. 9.1). People frequently assume that vegetation must be actively reintroduced to restore a site, but that is often not the case. In fact, *natural regeneration* can be a cost-effective option for recovery, particularly in sites that are lightly *degraded* and have nearby seed sources. One can assess the likely rate of natural recovery for a system by reviewing similar sites in the region or waiting for a few years to observe the rate of natural regeneration in the area to be restored (chap. 5).

If natural regeneration occurs too slowly to meet project objectives, methods to *assist regeneration* by enhancing the natural rate of vegetation

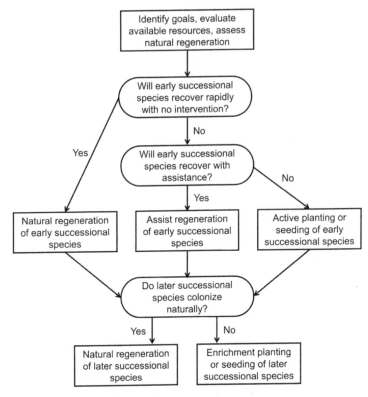

Figure 9.1. Decision tree for planning how and when to intervene to restore vegetation. Figure revised from Holl 2012.

recovery should be considered. As discussed in chapter 5, reinstating *disturbance regimes*, such as flooding or fire, can facilitate vegetation establishment in ecosystems adapted to those disturbances. Likewise, restoring *abiotic* conditions (chaps. 6 and 7) or controlling *invasive species* (chap. 8) often enhances *native* vegetation establishment. One way to increase the rate of natural regeneration is to create *microsites* that facilitate seedling establishment (chap. 6; Whisenant 1999). For example, Tongway and Ludwig (1996) introduced piles of dead branches in degraded shrublands in eastern Australia, which trapped *organic matter* and seeds, provided habitat for mammalian seed dispersers, and moderated microclimate conditions, resulting in enhanced seedling establishment and increased *spatial heterogeneity*. Assisting natural regeneration processes often requires less effort to restore a diverse vegetation community than actively reintroducing all the species. However, in highly degraded ecosystems, assisted regeneration may not result in the desired rate of recovery or species composition. If natural or assisted regeneration does not meet project goals, then *active restoration* (e.g., planting or seeding) will be needed to reintroduce desired plant species (see fig. 9.1). *Practitioners* must decide how extensively and when during the recovery process to plant. Many projects actively *revegetate* the entire project site, particularly when the soil is bare. Another approach is to introduce small patches or clusters of vegetation (Rey Benayas, Bullock, and Newton 2008; Corbin and Holl 2012) that serve as *nurse plants* to *facilitate* the establishment of other plant and animal species. In recovering forest systems, tree clusters attract seed-dispersing birds and shade out light-demanding grasses, which enhances seedling *recruitment* of a diverse suite of species and increases tree canopy cover (Corbin and Holl 2012). In arid and semiarid sites with stressful abiotic conditions, shrubs act as nurse plants by trapping seeds, increasing soil organic matter and nutrients, and reducing soil and air temperatures (Gomez-Aparicio 2009). Establishing patches of vegetation is less costly than planting or seeding an entire site and provides more spatial heterogeneity.

In systems in which early successional species naturally regenerate quickly, a potential method to accelerate the establishment of later successional species is *enrichment planting*. Once the initial colonizers have established and created favorable conditions for later successional species (e.g., a canopy providing shade in forested systems), practitioners can introduce species that would colonize slowly or not at all without intervention. Enrichment planting has been used in ecosystems ranging from prairies in the midwestern United States (Greene and Curtis 1953) to tropical forests (Rodrigues et al. 2009) to ensure that desired later-successional species establish.

Selecting Species to Plant

The species chosen for revegetation depend on the goals of the project and the resources available. Some projects focus intensive efforts on reintroducing one or a few *species of concern*, whereas projects aimed at restoring ecosystems and their processes typically reintroduce several plant species. Some ambitious and well-funded restoration projects, such as reclamation of aluminum mines in Australia (Koch 2007) and the Atlantic forest in Brazil (Rodrigues et al. 2009), aim to reintroduce more than 100 species, but they are exceptional cases. A more common approach is to revegetate with a subset of species that will facilitate the establishment of other organisms.

Table 9.1. Potential Characteristics to Consider in Selecting Plant Species for Restoration

Characteristic	Description and rationale
Growth rates	Rapid ground cover of herbaceous species to prevent erosion; fast biomass gain for carbon storage; rapidly forming tree canopy to shade out light-demanding, early successional vegetation; good to plant species with varying growth rates so that some establish quickly and others live longer
Growth form—e.g., herb, shrub, tree	Growth form(s) selected will affect vegetation structure and diversity
Tolerance of low-nutrient soils and nitrogen fixation	Able to grow and improve soil conditions in degraded sites
Tolerance of acidity, salinity, and toxic substances	Adapted to the specific soil chemistry in a degraded site and the salinity regime in coastal systems
Tolerance of stressful and changing climatic conditions	Tolerant of variable temperature and moisture conditions to be able to establish in degraded sites and survive in a changing climate
Traits that attract fauna	Fruits that attract seed-dispersing fauna, nectar sources, or species that provide habitat structure for fauna
Conservation concern	Species that are rare and the focus of conservation efforts
Likelihood to establish naturally	Plant species that are unlikely to colonize naturally
Feasible to collect and propagate	Increases cost efficiency and ease of restoration
Desirability as wood, as nontimber forest products, or for other economic or cultural reasons	Provides income, food, or other products, which increases the incentive for landowners to plant and maintain vegetation

For example, the *framework species method* involves planting ten to forty species that attract fauna and represent different growth rates and successional stages for tropical forest restoration (Goosem and Tucker 2013).

Selecting plant species for a restoration project requires weighing many different ecological and social criteria (table 9.1; Meli et al. 2014; Chechina and Hamann 2015). Typically, some planted species are tolerant of the stressful abiotic conditions typical of disturbed sites. *Nitrogen-fixing species* (chap. 6) are often included on nutrient-poor soils. Fast-growing, early successional species increase vegetation cover quickly and outcompete invasive species, but should facilitate the establishment of other plant species and fauna over the long term rather than inhibiting recovery of the site.

If later successional species are unlikely to colonize naturally, then the planting mix should include these species or enrichment planting should be done later (see fig. 9.1). Although forest restoration efforts typically focus on planting trees, they should include a variety of growth forms (e.g., understory species, epiphytes, vines) because there is no guarantee that all will colonize naturally. By default, species that are common and easier to propagate tend to dominate the suite of species planted (Brancalion et al. 2018; Lesage, Howard, and Holl 2018). Another important criterion to consider is whether species have cultural or economic value to *stakeholders*, which increases community engagement with restoration efforts (Meli et al. 2014).

Whether to include *nonnative*, noninvasive species as part of the initial planting mix is highly controversial. In highly degraded ecosystems, nonnative species may improve degraded abiotic conditions (Whisenant 1999; D'Antonio, August-Schmidt, and Fernandez-Going 2016). Sterile barley is often seeded in *riparian* restoration efforts in California because it provides rapid ground cover and erosion control, but does not reseed (Rein et al. 2007). In forest restoration, planting nonnative trees that provide a rapid canopy cover can shade out invasive grasses and other light-demanding ground cover species, attract seed-dispersing animals, and provide fuelwood, timber, or fruits to landowners (Feyera, Beck, and Lüttge 2002). Before nonnative species are included in a planting mix, however, serious consideration must be given to their potential for spread and long-term effects on the ecosystem, such as altering nutrient cycling, reducing water availability, or inhibiting the establishment of later-successional native species (D'Antonio, August-Schmidt, and Fernandez-Going 2016).

Vegetation Collection Guidelines

Seeds and seedlings of common species are often available from local nurseries, but obtaining the full suite of desired species or the preferred *ecotype* (i.e., locally adapted subpopulation of a species) often requires practitioners to collect seeds or other vegetative material from local sources. Table 9.2 summarizes recommendations for collecting seeds or plants to maintain *genetic diversity* and to minimize the effects of seed collections on source populations of rare species (Vitt et al. 2010; Maschinski and Haskins 2012). Recommendations include collecting from multiple plants across *environmental gradients*, from plants of different sizes, and over multiple flowering days, as well as clearly labeling collections with all relevant information (see table 9.2). It may be necessary to get a seed-collecting permit, depending on the rarity of the species and the landowner of the source

Table 9.2. Recommendations for Collecting Seed or Other Vegetative Material for Restoration

- Collect from a minimum of fifty plants to capture 95% of the genetic diversity.
- Collect across any obvious environmental gradients.
- Collect from both within the center of population density and from the periphery to ensure the greatest genetic diversity.
- Collect even the smallest plants because they may contain trait variations that would preadapt them to an alternate site.
- Collect at peak seed maturity or collect on multiple days.
- Collect from within the entire flower head and flowering branch.
- Collect no more than 10–20% available seeds on any given day.
- Individual collections should be stored separately with detailed collection information, including collectors' name, date of collection, locality information (GPS coordinates), and information that might help with habitat matching, such as soil type, terrain, and abundant associated plant species.

Source: Condensed from Vitt et al. 2010 and Maschinski and Haskins 2012.

populations. After collection, seed should be stored properly: protected from seed predators and fungi, at low humidity and temperature to reduce biological activity, and with an informative label.

A question that has long been discussed in the restoration literature and among practitioners is how locally to collect seed (Havens et al. 2015). In the past, the typical practice was to collect seeds as locally as possible because studies have shown that many species are adapted to local site conditions; hence, they have higher growth and survival rates when collected from sites nearby or from areas with similar abiotic conditions (Montalvo and Ellstrand 2000). Determining how locally to collect, however, is not an easy decision; it depends on the reproductive biology of a given species, the seed sources available, and the project goals. Some conifers may disperse pollen over hundreds of kilometers and hence have a similar genetic composition across a broad geographic range, whereas many small herbaceous plants may only exchange pollen or disperse seeds within a hundred meters of a given population. Guidelines for seed collecting zones are available for some widespread or economically valuable species (Bower, St. Clair, and Erickson 2014), but resource managers usually decide where to collect seeds based in large part on professional judgment and feasibility. If there are insufficient seed sources nearby, then they collect seeds from a broader area. When a project aims to restore a rare species with a few different populations, it is important to keep plant materials from those populations separated to prevent genetic mixing.

The issue of how to locally collect seed and other plant materials has become further complicated by anthropogenic climate change. Selecting ecotypes that will be able to withstand warmer temperatures, altered precipitation, and changing *inundation* patterns will become increasingly important for future success. Scientists and practitioners debate whether to collect seed as locally as possible, introduce species or ecotypes that are more likely to be adapted to future climatic conditions, or use a diversity of species and ecotypes and let site conditions sort out which are best adapted (Breed et al. 2013; Havens et al. 2015; Prober et al. 2019). An increasingly common approach is to increase "adaptive capacity" by introducing plants from a diverse set of seed sources to increase the likelihood that some will survive, regardless of the abiotic conditions at the start of the restoration project and in the future (Prober et al. 2019). Some projects try to select genetic material to match predicted future conditions, but that can be challenging given the uncertainty in climate models. Breed et al. (2013) provide a decision tree for how to source vegetative material based on the availability of climate models and biological information on the target species, but it remains difficult to give resource managers general rules of thumb on how locally to collect.

Plant Propagation Methods

Plants can be introduced to a site in various ways: as seeds, in topsoil, through vegetative cuttings, or as nursery-grown or wild collected plants of different sizes. Choosing among propagation methods depends on an understanding of the reproductive biology of target species, as well as the scale of and resources available for a project. Rieger, Stanley, and Traynor (2014) provide a thorough discussion of the advantages and considerations for each approach, briefly summarized in Table 9.3. Regardless of the source of plant material, practitioners should begin to plan revegetation projects a year or more in advance to either ensure that sufficient seeds or plants are available from commercial suppliers or to collect and grow plants for a project.

Seeds

Introducing vegetation by seeding can be cost-effective because it does not require nursery facilities or extensive labor to grow and plant seedlings. However, the drawback is that seed predation, low germination, and mortality of recently germinated seedlings due to desiccation and herbivory can result in low recruitment. For example, Holl (2002a) found that only 0.2 percent of seeds in abandoned tropical pastures survive to become eighteen-month-old seedlings. Thus, restoring with seed requires large

Table 9.3. Advantages and Considerations for Different Propagation Techniques

Technique	Advantages	Considerations
Direct seeding	• Cost-effective; can be done over large areas • Lower propagation and transport costs • Seeds may remain dormant until conditions are appropriate • Founding populations usually include more individuals and genetic diversity • Less likely to spread plant pathogens	• Potential for low establishment due to seed herbivory, erosion, or low germination • Establish and grow more slowly than seedlings • Large numbers of seeds needed • Seeds may require pretreatment • Need to be careful not to introduce weed seeds
Planting seedlings	• High establishment rate • Can inoculate mycorrhizae in the nursery • Larger when installed, so more likely to be able to outcompete invasive species	• Requires a greenhouse or shade house to grow seedlings • More expensive • Greenhouse-grown seedlings can experience transplant shock, particularly if outplanted to a site with stressful abiotic conditions • Well-fertilized plants may experience high herbivory
Cuttings/vegetative propagation	• Can cost less, depending on methods • Some species can be collected and immediately transplanted and so do not need greenhouse facilities	• Only works for species that readily grow roots from cuttings; may need to use rooting hormones • Low genetic diversity • Not always sufficient source material; can cause damage to plants
Salvage of vegetative material or soils	• Make use of soil and vegetation that might have been destroyed • Salvaging soil introduces a diverse microbial community and seed bank	• Often no source of material nearby without damaging other ecosystems • Soil may introduce unwanted species • Plants usually require immediate irrigation

quantities of seed relative to the number of plants desired. Nonetheless, seeding usually costs less than revegetating with nursery-grown plants. In central Brazil, up to 30 hectares of tropical forest can be seeded mechanically each day at a much lower cost than planting seedlings (Durigan, Guerin, and da Costa 2013).

Seeds can be purchased commercially or collected specifically for a given project. Although purchasing seed is easier logistically, it may be difficult to obtain seeds of all desired species and of locally collected ecotypes (Rieger, Stanley, and Traynor 2014). Ladouceur et al. (2018) found that only 39 percent of species of interest for restoring European grasslands were available commercially, and this percentage is certainly much lower for ecosystems with restoration initiatives that are not as developed. The process of manually collecting seeds and cleaning them to remove fruit pulp or other nonseed parts (such as pods or cones) can be extremely time-consuming. Furthermore, the amount of locally collected seed can be limited by the size of the source populations, especially for rare species. The quantity of seed can be increased by growing the initial collections of a species in a greenhouse or on seed farms in the field and then harvesting the seed. As restoration becomes more common, the quantity and variety of commercially available seed is growing. For example, in the Brazilian Atlantic forest, training a seed collector network to provide sufficient seeds has been essential to increase the spatial scale and tree diversity of restoration projects (Atlantic Forest case study; Brancalion, Viani, Aronson, et al. 2012).

It is important to know the *seed viability* (germination potential) of species used in restoration. This information should be provided for commercially purchased seed, but needs to be tested if seeds are collected by project staff. The ideal seeding rate is highly site and species specific and should be determined based on seed viability and pilot studies or previous restoration projects in the focal ecosystem. Seed should be kept as pure as possible to avoid introducing seeds of invasive species.

Information on seed biology and germination cues is crucial to selecting methods that will maximize revegetation success. Most seeds from temperate terrestrial systems can be dried to less than 5 percent of their fresh weight and stored for many years without losing their viability. Some species germinate readily, but others have evolved *seed dormancy* mechanisms to ensure that they do not germinate when environmental conditions are unfavorable. Such species often require complex germination cues to break their dormancy. Many species from temperate zones have evolved to germinate in the spring and therefore need to be *stratified* (exposed to a period of cold temperatures) before germination will occur. Early work on prairie restoration at the University of Wisconsin Arboretum found that stratification was necessary for the germination of 39 percent of the species and improved the germination rates of an additional 34 percent of species (Greene and Curtis 1950). Day length (i.e., a cycle of light and dark conditions) and exposure to direct sunlight are important triggers for the germination of some species. Other species have thick, hard seed coats, which enable them to withstand dispersal in the stomachs of animals or to allow them to persist in the *soil seed bank* for many years. For example, some dune species have hard seed coats that prevent germination until they are worn down by shifting sand; these species need to be *scarified*, meaning that their seed coats must be worn down through chemical (e.g., acid) or mechanical (e.g., sandpaper) means so that they can absorb water and germinate. For seeds that have evolved under a fire regime, exposure to short periods of high temperature or the chemicals from wood smoke may stimulate germination.

In contrast, some plants have seeds with high moisture content (greater than 50 percent) that lose viability if they are dried and stored. These plants are often found in tropical rain forests and aquatic environments where moisture is not a strong constraint to establishment. The seeds germinate readily but cannot be stored for future use. Given that many plants do not set seed every year, it can be a challenge to obtain a sufficient seed supply for a restoration project. Some forest nurseries in the tropics sow the species in a low-nutrient soil and in shaded conditions, reducing their growth and creating a seedling bank for use in the future.

Various methods are used to distribute seed in a restoration site. Small seeds can be dispersed mechanically over large areas by hydroseeding or drill seeding. *Hydroseeding* involves mixing a slurry of seeds, nutrients, wood mulch, and water and then spraying the mixture; the technique is particularly useful on uneven terrain. *Drill seeding*, using tractors with specialized attachments or hand-held drill seeders, buries seeds right below the soil surface, which improves seed-soil contact and in turn reduces seed desiccation and loss to erosion and predation. Seed-sowing machines used in crop production can be adapted for direct seeding in restoration projects (Durigan, Guerin, and da Costa 2013). When seeds are spread directly on the soil surface, they often germinate readily, but there is higher potential for losses to seed predation and desiccation. In rare cases, helicopters or planes have dispersed seeds and fertilizer over large areas or remote, hard-to-access landscapes (Elliott 2016).

Vegetative Propagation

Some species are adapted to resprout after disturbance or to spread vegetatively. In these situations, *vegetative propagation* (growing new plants by cutting branches, stems, or roots from other plants) is a good strategy. For example, several species used to revegetate aluminum mines along the western coast of Australia can only be propagated through cuttings or by culturing plants from tissues in the lab (Koch 2007). In some cases, vegetative material is directly planted at the restoration sites, and in other cases, cuttings are put in a greenhouse until they develop root systems. It is common to take a small section of stem and leaves from wetland grasses, rushes, and sedges and then immediately transplant the section to a restored site where high moisture conditions favor growth of new roots. In tropical regions where species have evolved to resprout after hurricanes and floods, some tree species can be propagated by cutting a 1- to 2-meter branch from a tree and placing the branch directly into the ground, where it will root and grow. A similar approach is used for willows (*Salix* spp.) in temperate riparian forests, although only some tree species can grow new roots and establish in this way. As vegetative material produces clones, it is important to take cuttings from enough plants to ensure high genetic diversity and minimize damage to individual plants.

Nursery-Grown Plants

It is common to grow plants in a greenhouse or purchase commercially grown nursery seedlings and outplant them on the restoration site. This approach reduces losses to seed predation and the desiccation of recently established seedlings and optimizes the use of collected seeds, which can be too costly or difficult to obtain in quantities sufficient for direct seeding. Nursery-grown seedlings can be produced from collected seeds, plant cuttings, or seedlings collected in the field and then held temporarily in the greenhouse to acclimatize to the high light conditions typical of restoration sites.

Greenhouse facilities vary from climatically controlled greenhouses to rudimentary outdoor facilities that provide only shade and water for seedlings. When seedlings are grown in greenhouses with temperature control, shade, regular watering, and fertilization, hardening seedlings prior to planting may decrease transplant shock and increase survival in the field. *Hardening* is the process of preparing plants for the stress of the natural environment by gradually subjecting them to increasingly more field-realistic levels of sunlight, moisture, and temperature before they are transplanted.

Growing and planting seedlings typically costs more than seeding, but it results in much higher plant survival and species diversity because many species do not establish well from seeds in field conditions. Larger plants have greater survival but are costlier to grow and outplant, so the smallest possible size that will survive in the field should be used (Rieger, Stanley, and Traynor 2014).

Transplanting Soil or Plants

Collecting topsoil, seedlings, or hay from existing habitat is another potential source of plant material. The most desirable method of acquiring such material is to salvage soil or plants from an area prior to disturbance (e.g., mining) or from a nearby site that is slated to be cleared (e.g., building construction site). Practitioners should only remove a small amount of soil or plants from intact ecosystems and carefully consider the effect on the source site.

Topsoil can be a source for seeds of native species, as well as microbial communities. For example, at least twenty-eight plant species were successful introduced to reclaimed aluminum mines in Australia by either *translocating* or *stockpiling* and replacing soil (chap. 7; Koch 2007). Ferren et al. (1998) report on a successful effort to reintroduce plant and microbial communities through translocating small amounts of soil between remnant and restored vernal pool wetlands in Santa Barbara, California.

In some forest systems, more tree seedlings establish below a single mother tree than will ever survive. These seedlings and saplings can be collected and transplanted to restoration sites. However, they may suffer high mortality if immediately transplanted from a shady understory to an open area with high light and temperature conditions, so they should be hardened in a nursery prior to planting. In European grasslands and wet meadows dominated by native species, hay is mowed, collected, and transferred to restoration sites as a source of seeds (Klimkowska et al. 2010) and can serve to reintroduce microorganisms. When using hay or soil to revegetate an area, it is critical to consider whether any undesirable plants, animals, or microorganisms will be transferred along with the focal organisms.

Enhancing Plant Survival and Growth

Careful site preparation prior to planting, to overcome site limitations such as nutrients, soil moisture, and *competition* with invasive species, enhances revegetation success. Likewise, managing the site for an extended period following project implementation increases the likelihood of achieving project *objectives*. The specific methods used to improve vegetation survival and growth are ecosystem and site specific. For example, in tidal wetland or seagrass systems, it may be necessary to secure plants so that they are not washed away by waves. I briefly discuss a number of common approaches below.

Site Preparation and Weed Control

Where the *landform* and soils have been highly altered, revegetation efforts are unlikely to succeed until the *topography* and *hydrologic regime* of the site are restored (chap. 6). In sites dominated by invasive or otherwise undesirable species that are likely to outcompete the desired native species, it is critical to implement control methods well prior to seeding or planting. Various methods are used, such as herbicides, mowing or clipping, and applying wood or cardboard mulch to inhibit germination from the seed bank (chap. 8).

Tailoring Planting Species Mix to Heterogeneous Site Conditions

The same mix of plant species is often seeded or planted across an entire site. Planting success is higher, however, when restoration practitioners conduct detailed site assessments to better characterize small-scale variation in soil type and depth, depth to groundwater, and topographic position before developing detailed revegetation plans. Such assessments enable them to choose the species that are best adapted to localized conditions and increase habitat heterogeneity. In wetlands, that means selecting the species planted to match their topographic position and tolerance of inundation. Along the Sacramento River, for example, land managers plant grasses and shrubs on sandy shallow soils and riparian forest trees where there are deeper soils to support them (Sacramento River case study). Tailoring the planting mix to microsite conditions not only increases plant survival, but also enhances the restoration of ecosystem processes, such as nutrient and hydrological cycling (McCallum et al. 2018).

Timing

To enhance the likelihood of project success, revegetation efforts should ideally be timed to coincide with the natural establishment and growth cycle of a given ecosystem. In most ecosystems, it means planting at the beginning of the rainy season, which reduces the need for irrigation. In

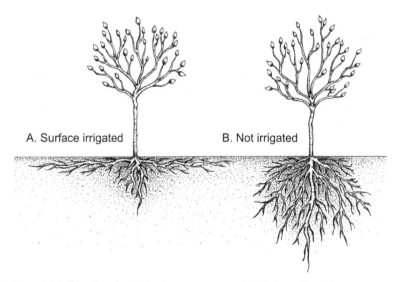

A. Surface irrigated   B. Not irrigated

Figure 9.2. Effect of surface irrigation on root growth. (A) A small tree that was surface irrigated, stimulating growth of roots near the soil surface. (B) A small tree with deeper roots typical of plants that are not irrigated or irrigated below ground, which enables them to reach deeper water sources and survive dry periods. Drawing by A. M. Baca.

ecosystems in which seeds have adapted to germinate after a winter cold period, seeding in the fall allows the seeds to stratify naturally and enhances germination in the spring. In riparian systems, plants should be seeded or planted with enough time for them to establish before the flood season, thus reducing the likelihood that they are washed downstream.

Water Availability

In arid systems and areas with seasonal rainfall, water availability is often a major factor influencing seedling survival and growth. Irrigating for the first one or few years often increases plant survival in restoration projects where water is limited and project funding is sufficient, but it also increases restoration costs and requires a nearby water source. Moreover, irrigation should be done judiciously so that irrigated plants develop the deep root systems required to access groundwater over the long term (fig. 9.2); tree seedlings that receive extensive surface irrigation are less likely to survive dry conditions and are more susceptible to windthrow. Many other strategies are used in arid systems to increase soil moisture and to enhance survival of plants (Bainbridge 2012). Different types of mulch help maintain soil moisture, planting seedlings in *microcatchments* increases localized soil moisture, and synthetic gels absorb and then slowly release moisture after rainfall events (Hüttermann, Orikiriza, and Agaba 2009). In some deserts and drylands, plants are grown in tall pots and planted in deep holes to encourage deep root growth (Bainbridge 2012).

Nutrient Availability and Introducing Beneficial Microorganisms

Highly degraded sites often have low organic matter and soil nutrients (chap. 7). Fertilizing plants at the time of planting or once they have regrown the fine roots that were damaged by transplanting can enhance plant survival and growth. Slow-release fertilizers provide plants with nutrients over a few weeks or months and minimize runoff of excess nutrients. In most cases, ongoing nutrient inputs are not advisable because fertilizing increases costs and because fertilizing plants in naturally low-nutrient soils favors invasive species that more rapidly take advantage of the additional nutrients. Furthermore, elevated nutrient concentrations in plant leaves can increase herbivory.

Reintroducing *mutualistic* microorganisms can improve nutrient availability (chap. 7). Mycorrhizae often colonize restoration sites naturally and enhance plant growth by aiding with the uptake of water and nutrients. For some plant species that are dependent on mycorrhizal associations, actively reintroducing the mycorrhizae can be critical for successful establishment of the plants. *Mycorrhizae* and other microbes can be introduced into the soil of individual nursery-grown plants or directly to the restoration site.

Herbivory

Seed predation and plant herbivory can be major factors limiting plant establishment, particularly in systems where natural predators are absent and therefore do not control insect or vertebrate herbivore populations. Reintroducing natural predators to regulate herbivore populations is the most sustainable long-term solution, but may not be possible (e.g., reintroducing native mammalian predators to urban areas to control rabbits). Species that have natural chemical or physical defenses against herbivory should be planted preferentially where herbivory is a problem. Other approaches include fencing entire sites and caging individual plants to exclude vertebrate herbivores. Insecticides can be used to reduce herbivory, but they have potential negative effects on other organisms in the ecosystem and humans and may be prohibited. Where seed predation is high, covering the seeds with a thin layer of soil or seeding as close as possible to the time of germination can reduce predation.

Ongoing Adaptive Management

Ongoing *adaptive management* enhances the survival and growth of planted vegetation and facilitates the establishment of other species. Adaptive management may take the form of short-term fertilization or irrigation. It typically includes some form of controlling competitive and undesired plants through any of the many methods discussed previously (chap. 8), such as manual removal, targeted use of herbicides, or restoring a disturbance regime that favors the desired species. These management actions may be required for a few years following planting or may be needed on an ongoing basis over many years. It is important to *monitor* whether restoration objectives are being met and then adjust ongoing management actions accordingly (chap. 3).

Recommended Reading

Bainbridge, David A. 2012. *A Guide for Desert and Dryland Restoration: New Hope for Arid Lands*. Washington, DC: Island Press.

    Describes detailed methods for restoring arid lands.

Havens, Kayri, Pati Vitt, Shannon Still, Andrea T. Kramer, Jeremie B. Fant, and Katherine Schatz. 2015. "Seed sourcing for restoration in an era of climate change." *Natural Areas Journal* 35:122–33.

    Provides a thoughtful summary of options for sourcing plant material in a changing climate.

Maschinski, Joyce, and Kristin E. Haskins. 2012. *Plant Reintroduction in a Changing Climate: Promises and Perils*. Washington, DC: Island Press.

    Reviews rare plant reintroduction projects and practices, as well as lessons learned for reintroductions to preserve species threatened by climate change.

Rieger, John, John Stanley, and Ray Traynor. 2014. *Project Planning and Management for Ecological Restoration*. Washington, DC: Island Press.

    Thoroughly discusses the many detailed decisions required to plan and implement a terrestrial restoration project.

Rodrigues, Ricardo R., Renato A. F. Lima, Sergius Gandolfi, and André G. Nave. 2009. "On the restoration of high diversity forests: 30 years of experience in the Brazilian Atlantic Forest." *Biological Conservation* 142:1242–51.

    Describes detailed methods for restoring a diverse plant community in the Brazilian Atlantic forest.

10

Fauna

Many restoration projects focus on restoring *abiotic* conditions, *ecosystem services,* and plant *community composition,* overlooking the important roles played by a wide diversity of fauna in the *recovery* process. Faunal conservation and restoration efforts often focus on large, charismatic, vertebrate species or groups of vertebrates, such as birds, mammals, fish, or amphibians. There are, however, millions of species of invertebrate animals, such as insects, spiders, mollusks, crustaceans, and a host of soil fauna that play important ecological roles.

Plant-animal interactions strongly influence the success or failure of restoration efforts (McAlpine et al. 2016). Several faunal groups are important seed dispersers and pollinators and are therefore critical for the establishment and reproduction of a diversity of plants. For example, the reintroduction of giant tortoises on Española Island in the Galapagos re-initiated seed dispersal and seedling *recruitment* of the *endangered* tree cactus (*Opuntia megasperma* var. *megasperma*; Galapagos Tortoise case study). Kaiser-Bunbury et al. (2017) found that removing invasive shrubs in the Seychelles increased pollinator richness and number of visits to several endemic plant species, which in turn increased fruit set. Soil fauna, such as earthworms and springtails, along with burrowing mammals, improve soil structure and increase water infiltration (chap. 6). Coral reefs consist of polyps (small invertebrate fauna related to sea anemones) that form a *mutualism* with zooxanthellae (single-celled organisms that photosynthesize);

the polyps form the large coral colonies that provide habitat for numerous other marine organisms.

Some faunal restoration projects focus on an individual species of conservation concern, such as a charismatic vertebrate, whereas other projects concentrate on restoring habitat characteristics for target faunal groups. Faunal restoration efforts generally use one or more of three general approaches, which I discuss in this chapter: (1) reducing the specific drivers of faunal population decline, (2) improving habitat quantity and quality, or (3) reintroducing individual faunal species. The first two strategies should be employed prior to reintroducing species, but that often does not happen.

Reducing Drivers of Faunal Decline

A first step in restoring an ecosystem and the fauna that live there is to resolve or at least reduce the initial cause of *degradation* (chap. 1). In some cases, this step alone is sufficient to drive a rebound in faunal population size. Land conversion is a major cause of faunal declines, but there are several others, such as overharvesting, predation by and *competition* with *invasive species*, and toxic chemicals.

Overhunting and overfishing have caused the decline of numerous faunal species around the globe. Although many species continue to be overexploited, there are some success stories where *management* has allowed populations to recover. In western Greenland, populations of common eiders (*Somateria mollissima*), a large sea duck, increased by 212 percent over seven years after the length of the harvest season was reduced (Merkel 2010). In recent years, an increasing number of marine protected areas and no-take zones for fishing have been established, which help species recover for conservation purposes and enhance fishing stocks over the long term. Thaman et al. (2017) compiled *traditional ecological knowledge* of historic species distributions and more recent field surveys to show that more than three hundred mollusk species were either seen for the first time in forty years or increased in population size after the establishment of a locally managed marine protected area in Fiji.

Another frequent cause of mortality of fauna populations, particularly on islands, is *invasive nonnative* predators (chap. 8). For example, the introduction of the brown tree snake (*Boiga irregularis*) played a primary role in the *extinction* or dramatic population declines of many small bird, mammal, and lizard species in Guam and other Pacific Islands (Fritts and Rodda 1998). If species have already gone extinct, then removing nonnative predators will not reverse the trend, but several examples show that target species' populations rebound after the *eradication* or reduction of predatory species. For example, the nongovernmental organization Island Conservation has removed invasive cats, foxes, goats, and other introduced nonnative animals from more than sixty islands worldwide, which has benefited the populations of numerous *native* seabirds, lizards, plants, and small mammals (Island Conservation 2017). Likewise, the removal of invasive predatory species from lakes through fishing or chemical means has benefited native fish species. For example, heavy fishing of the invasive Nile perch (*Lates niloticus*) in Lake Victoria in Africa has led to population increases for some of the hundreds of fish species that suffered dramatic declines following the perch's introduction (Witte et al. 2000).

Toxic chemicals in the environment negatively affect both terrestrial and aquatic fauna. When pollutants are a major driver of faunal population declines, the first step toward restoring these habitats is to reduce chemical inputs (chap. 7). For example, reducing the sulfate emissions in Norway decreased river aluminum concentrations and acidity, and the improved water quality triggered increases in sensitive aquatic invertebrate populations (Raddum, Fjellheim, and Skjelkvåle 2001). Vulture populations declined by 99.9 percent between 1992 and 2007 on the Indian subcontinent due to diflofenac, an anti-inflammatory drug used by veterinarians, which poisons vultures when they consume carcasses of dead animals treated with the drug (Cuthbert et al. 2014). The vulture decline led to a buildup of animal carcasses, particularly of the sixty-five million cows that die each year in India but are not consumed by humans for religious regions, and an increase in rat populations. Diflofenac has since been banned in India, Nepal, Bangladesh, and Pakistan, which has helped stabilize vulture populations in some regions, but the drug is still dispensed illegally.

Restoring Habitat Structure

The most common faunal restoration strategy is to restore habitat quality to encourage specific species or groups of fauna to recolonize a site. This approach, which has been variably successful, assumes that if certain habitat characteristics are restored (table 10.1), then fauna will recolonize. The failures in part stem from poor knowledge of the complex habitat needs of both individual species and faunal groups, as well as landscape-scale factors such as small habitat patch size and lack of connectivity with source populations (chap. 5).

Restoring habitat requires providing all the resources and abiotic conditions needed by an organism. Many species need heterogeneous habitat that provides different resources, depending on the time of the year or the stage of their life cycle (Morrison 2009). For example, butterflies need host

Table 10.1. Habitat Needs for Specific Faunal Groups[1]

Habitat variable	Mammals	Birds	Amphibians & Reptiles	Terrestrial Insects	Soil fauna	Fishes	Aquatic invertebrates
Abiotic							
Soil, air, or water temperature	★	★	★	★	★	★	★
Rocks for shade	★		★		★		
Soil or water quality—nutrient, toxics, pH	★	★	★	★	★	★	★
Soil texture, compaction, organic matter					★		
Soil moisture			★	★	★		
Water salinity	A[2]	A				★	★
Water turbidity	A	A	★			★	★
Water level/flow rate	A	A	★	★		★	★
Streambed substrate			★			★	★
Vegetation							
Canopy cover—tree or shrub	★	★	★	★	★	★	★
Ground cover—herbaceous or litter	★	★	★	★	★		
Food sources—fruit, leaves, nectar	★	★	★	★	★	★	★
Individual host plant species				★			
Brush piles (terrestrial), large woody debris (rivers) for shade and nesting habitat	★	★	★			★	
Species interactions							
Availability of prey	★	★	★	★	★	★	★
Invasive species control—predation or competition	★	★	★	★		★	★
Disease control	★	★	★			★	
Presence of mutualists	★	★	★	★	★	★	★
Landscape							
Presence of water sources	★	★	★	★		★	★
Nesting/spawning habitat	★	★	★	★		★	
Connectivity/proximity to source population	★	★	★	★	★	★	★
Habitat patch size—sufficient home range and population size	★	★	★			★	
Habitat heterogeneity—multiple habitat types for different life stages or uses	★	★	★	★	★	★	★

[1]List includes most important variables for different groups.

[2]A = aquatic species only.

plants for caterpillars, as well as nectar sources and open areas to thermoregulate for adults. The northern Mexican garter snake (*Thamnophis eques megalops*) uses wetland edge habitat during the warm portion of the year, but spends more time in rocky upland habitats during the winter when it is less active (Sprague and Bateman 2018). Many aquatic organisms use different habitats for the juvenile and adult stages of their life cycle. Nonetheless, restoration efforts tend to focus on restoring the habitat needed by one portion of a faunal species' life cycle, but coordinated efforts are needed to restore habitats that provide for all stages of a species' life history.

Most often, restoration projects focus on restoring specific habitat characteristics considered important for specific groups or species (see table 10.1). For example, a complex vertical and horizontal structure of trees and shrubs provides habitat for a diverse suite of birds, whereas many insect species need specific larval host plants. River restoration often focuses on improving fish habitat by removing barriers to movement, creating pools as refuges from high temperatures, and restoring *riparian* vegetation. Installing logs and woody debris in both terrestrial and aquatic systems provides shade and protected areas for nesting of many faunal species. In coastal reef systems, various structures (e.g., cement blocks, oyster shells, rocks) can be set out to enhance the amount of hard substrate available for settling of the larvae of oysters and coral polyps (Ferrario et al. 2014).

Whereas most habitat-focused efforts aim to restore groups of fauna (e.g., frugivorous birds, anadromous fish), some projects aim to restore habitat quality for individual species. Considerable research has focused on restoring habitat for the red-cockaded woodpecker (*Picoides borealis*) in the longleaf pine forests in the southeastern United States. Methods include restoring an appropriate vegetative structure by thinning vegetation or reinstating a fire regime, as well as creating nest cavities and protecting them from being enlarged by other woodpecker species (Conner, Rudolph, and Walters 2001). Whereas these efforts focus on the red-cockaded woodpecker, they benefit other pine-grassland species, as well as some neotropical migrant birds (Wilson, Masters, and Buckenhofer 1995). In another example, the number of rhinoceros auklets (*Cerorhinca monocerata*), a seabird that lives on small islands off the west coast of North America, increased when ceramic nesting structures were installed to protect their underground nests from trampling by sea lions (*Zalophus californianus*) (Beck et al. 2015; see http://oikonos.org/seabird-nests/ for photos).

Restoring high-quality habitat is a necessary but insufficient approach to restoring faunal species. Fauna will only colonize a restoration site if there is a source population within the dispersal distance of the species and there are no barriers to dispersal. Strategies to enhance faunal movement include locating restoration sites near source populations, restoring *ecological corridors*, and managing the intervening land to increase *landscape connectivity* (chap. 5; Morrison 2009; McAlpine et al. 2016).

For example, eastern collared lizards (*Crotaphytus collaris collaris*) have a *metapopulation* structure and live in open, rocky habitat distributed patchily among a woodland matrix in the northeastern Ozarks, Missouri, but translocated populations were not colonizing suitable habitat approximately 50 meters away because fire suppression had resulted in a dense, woody understory (Templeton, Brazeal, and Neuwald 2011). After the forests were thinned using controlled burns, the lizards moved between the different habitat patches, reconnecting the population. As would be expected, habitat restoration for the southwestern willow flycatcher (*Empidonax traillii extimus*), an endangered bird native to the southwestern United States, has been more successful in sites nearer to source populations (Tamarisk Removal case study).

If barriers to population movement cannot be removed, restoration efforts should aim to provide alternative routes. For example, *fish ladders* and bypass channels may enable fish species that migrate upstream to pass by dams to spawn. In terrestrial systems, underpasses or overpasses have been installed to allow wildlife to cross roads safely (Taylor and Goldingay 2012; see book website for photos of structures to enhance faunal movement).

Moreover, larger faunal species need a large enough habitat area to ensure viable populations (chap. 5). For species that have large home ranges or for which a single restoration site only provides part of their habitat needs, restoration and land management should be coordinated across multiple landowners to be successful. For example, along the southern coast of Oregon, the Coos Watershed Association has coordinated restoration and applied research among private landowners, government agencies, and scientists throughout 2,800 hectares and 88 kilometers of streams to improve habitat quality for several fish *species of concern* (Wright and Souder 2018). Similarly, the US Natural Resources Conservation service has proactively worked with more than thirteen hundred ranches on more than 5 million hectares (an area the size of Costa Rica) in the western United States to improve habitat for sage grouse (*Centrocercus* spp.) through fire management, invasive species control, and habitat restoration in an effort to prevent the listing of two sage grouse species as endangered (US Department of Agriculture 2016).

Faunal species with large home ranges often cross the borders of countries, so successful restoration requires coordination across political boundaries, which is difficult (Kark et al. 2015). Governmental and nongovernmental partners in India and Nepal have been working on a large-scale strategy to conserve and restore landscape connectivity for tiger (*Panthera tigris*) populations that cross their borders (Wikramanayake et al. 2010). Restoring long-distance migratory species is a particular challenge because their home habitats are distant from one another. Seventy-seven countries and the European Union are part of the Agreement on Conservation of African-Eurasian Migratory Waterbirds, which aims to coordinate efforts to protect and restore habitat for 255 wetland-dependent migratory birds (Kark et al. 2015).

Single Species Reintroductions

Another common approach to faunal restoration in both terrestrial and aquatic systems is *reintroduction* or translocation of individual species to either improve their conservation status or to promote processes for ecosystem recovery, also referred to as *rewilding* (Seddon et al. 2014; Swan et al. 2016). Most often, they are species at risk of extinction or charismatic large animals that have substantial impact on the food chain or other key ecological processes. For example, reintroduction of the gray wolf (*Canis lupus*) in the greater Yellowstone ecosystem in the northwestern United States has had complex effects on various animal populations, vegetation dynamics, and river channel morphology. The wolves reduced elk grazing on riparian shrubs and trees (Beschta and Ripple 2016; East 2017), and the vegetation regrowth has favored beaver populations, which alter river channel meandering through their dam building.

Some animals are reintroduced from *captive breeding* efforts, whereby individuals are bred and their offspring are raised in captivity prior to release (Morrison 2009). This approach requires extensive knowledge about animal husbandry techniques. It also involves training the animals in necessary behaviors, such as hunting and predator avoidance. For example, prerelease training of captively bred juvenile black-tailed prairie dogs (*Cynomys ludovicianus*) to sound an alert call in response to predators increased their postrelease survival (Shier and Owings 2006).

In other cases, species are *translocated*, or moved from existing populations to new sites, which tends to be more successful than reintroduction from captive breeding because the animals have learned behaviors necessary to survive in the wild. For example, between 1997 and 2006, $3.5 million was spent translocating 218 Canada lynx (*Lynx canadensis*), a medium-size cat species, from populations in Canada and northern Wyoming to Colorado. In 2013, the Colorado population was estimated to be two hundred to three hundred individuals, which included third-generation lynx. When translocating species, it is important to consider the effects of removing individuals from the donor population. Another concern is moving animals sufficiently far enough away so that they do not return to their source populations (Destro, De Marco, and Terribile 2018). In some cases, animals are released into the wild with no subsequent human support (*hard release*), but more often, *soft release* is used. With soft releases, the animals are provided with food for a limited amount of time to allow them to transition, which has been shown to increase the success of reintroduction efforts (Fischer and Lindenmayer 2000). In some cases, animals that were recovered from the illegal pet trade, such as primates, parrots, and reptiles, are reintroduced, particularly if they were originally raised in the wild and the area from which they were collected is known (Banes, Galdikas, and Vigilant 2016).

Typically, reintroduction efforts are undertaken once populations are already quite small, so it is important to consider how to establish a population with enough *genetic diversity* to minimize *inbreeding depression* and to maintain a viable population over the long term (chap. 5; Galapagos Tortoise case study). Some translocation efforts aim to increase the population size or genetic diversity of existing populations rather than establish new populations (Corlett 2016). For example, several female Texas puma (*Puma concolor stanleyana*) were translocated to Florida to increase genetic variation and reduce inbreeding depression in the critically endangered Florida panther (*Puma concolor coryi*) (Hedrick and Fredrickson 2010).

Faunal reintroduction efforts are often quite costly. It is not unusual for costs to range from $10,000 to $20,000 per animal, such as in the Canada lynx example above. Nonetheless, many reintroduction efforts have been unsuccessful for a number of reasons (Fischer and Lindenmayer 2000; Destro, De Marco, and Terribile 2018).

First, in some cases, such as for the Asian vultures discussed earlier, the drivers of the original population decline have not been fully resolved (Fischer and Lindenmayer 2000; Destro, De Marco, and Terribile 2018) or animals are reintroduced into low-quality habitat. For example, efforts to translocate the giant weta (*Deinacrida mahoenui*), a flightless cricket that is native to New Zealand and grows up to 10 centimeters long, have had mixed outcomes (Watts and Thornburrow 2009). Success has largely been determined by whether invasive rodent predators were eradicated prior to weta reintroduction, because invasive predators are the primary cause of the weta's population decline. In Canada, the United States, and several European countries, large populations of the peregrine falcon (*Falco*

Table 10.2. Factors Increasing Success of Faunal Reintroduction Efforts

- Cause of population decline removed or reduced
- Larger number of individuals introduced
- Appropriate sex ratio of introduced animals
- Source population wild rather than captive-bred
- Soft releases more successful than hard releases
- Release into core area of population
- Herbivores more successful than carnivores
- Long-term commitment to reintroduction effort
  o Releases in multiple years
  o Ongoing monitoring and management

Sources: Fischer and Lindenmayer 2000; Morrison 2009; Destro, De Marco, and Terribile 2018.

peregrinus) were restored by reintroducing captively bred birds. These efforts were successful because the main cause of decline was removed, namely pesticides that reduced eggshell thickness, and because peregrine falcons can live in urban landscapes where they nest on roofs of tall buildings and under bridges.

Second, many faunal reintroduction efforts fail due to insufficient knowledge about the biology of the *focal species*. In early attempts to reintroduce the endangered Attwater's prairie chicken (*Tympanuchus cupido attwateri*) in Texas, the birds starved despite sufficient food availability. Only then did scientists realize that the prairie chickens had been raised on pellets and had not developed the full complement of microbes needed to digest their typical diet of vegetation, seeds, and insects (Griffin 1998). Therefore, it is important to include research and *monitoring* as parts of faunal reintroduction projects. Systematic reviews have highlighted several other factors that increase the success of faunal reintroduction, such as reintroducing into the core rather than the periphery of a species range, translocating species from a wild population rather than reintroducing them from captive breeding, and providing long-term management (table 10.2; Fischer and Lindenmayer 2000; Morrison 2009).

Faunal reintroduction, particularly of predators with large home ranges, can be controversial because of fear of harm to both people and domestic animals. Hence, a critical component of such efforts is understanding local knowledge and opinions about animals (e.g., Maheshwari, Midha, and Cherukupalli 2014; Lopes-Fernandes, Espírito-Santo, and Frazão-Moreira 2018) so that educational and other programs are designed to increase public acceptance. For example, several carnivore reintroduction programs offer financial or material compensation to ranchers who lose livestock to predators; such programs have met with mixed success, however, and need to be tailored to local communities (Maheshwari, Midha, and Cherukupalli 2014).

Improving Faunal Restoration Efforts

Despite the strong linkages between faunal and plant recovery, integration of animal and plant restoration is the exception rather than the rule (McAlpine et al. 2016), which is a concern given the numerous plant-animal interactions that affect ecosystem recovery. Going forward, restoration plans should consider interactions between abiotic conditions, microbes, plants, and animals at a sufficient scale for all species so as to improve the likelihood of restoration success (McAlpine et al. 2016; Hale et al. 2019).

Monitoring the success of faunal restoration is challenging. First, animals are mobile and often secretive, so extensive observation to determine whether they are using restored habitat may be required. Second, detailed observations are needed to determine how animals are using the restored habitat. Are they primarily living in adjacent habitat and just passing through the site quickly, or do restored sites provide all the resources needed for them to reproduce and establish a viable population over the long term? Hale et al. (2019) report that only 11 percent of faunal restoration studies measure *parameters* related to *fitness* (e.g., breeding activities, reproduction, or survival). A large concern is that restored sites might serve as an *ecological trap* where fauna are attracted to the habitat but where overall habitat quality is poor so that it reduces population fitness. For example, installing perching structures for lizards may increase their risk of predation by avian predators (Hawlena et al. 2010). Although demonstrated examples of restored habitats reducing population viability are few, long-term monitoring and more observations of animal behavior and reproduction in restored sites are needed to better evaluate the habitat quality and identify critical resources to improve future faunal restoration efforts (Lindell 2008; Hale et al. 2019).

Another difference between animal and plant restoration is animal sociality, which can strongly influence restoration success. For example, black-tailed prairie dogs are highly social and live in family groups. Although most prairie dog translocations have been unsuccessful, translocating prairie dogs as family groups increased survival fivefold (Shier 2006). A seabird colony of more than three hundred common murres (*Uria aalge*) was *extirpated* from a rock outcrop along the northern California coast following an oil spill in the early 1980s, and the murres did not recolonize the area for over a decade. In 1996, murre decoys were placed on the island and their vocalizations played loudly from speakers; the seabirds returned to the rock the same day and have remained ever since (US Fish and Wildlife Service 2012). These examples illustrate the importance of considering faunal social interactions in order to design successful faunal restoration strategies.

Recommended Reading

Corlett, Richard T. 2016. "Restoration, reintroduction, and rewilding in a changing world." *Trends in Ecology and Evolution* 31:453–62.
Reviews the controversies of reintroducting faunal populations that are currently living and that have previously gone extinct.

Lindell, Catherine A. 2008. "The value of animal behavior in evaluations of restoration success." *Restoration Ecology* 16:197–203.
Discusses why it is important to monitor animal behavior and habitat usage in restoration projects.

Morrison, Michael J. 2009. *Restoring Wildlife*. Washington, DC: Island Press.
Provides a thorough review of how to restore and monitor habitat for wildlife.

11

Legislation

Regardless of the many different motivations to undertake ecological restoration (chap. 2), most projects occur within a complex regulatory context. Regulations include laws that require restoration following certain environmentally degrading actions, as well as incentives for voluntary restoration and international agreements (Mansourian 2017). They not only provide a driver and guide for restoration, but also serve to institutionalize an obligation to compensate for environmental *degradation* as a societal norm (Telesetsky, Cliquet, and Akhtar-Khavari 2017). Regulations apply to many actors and in a variety of contexts; some apply only to governmental actors, others hold corporations and individuals responsible for activities that cause environmental damage, and even individuals or members of nongovernmental organizations who voluntarily undertake restoration activities are often subject to regulations.

These regulations not only mandate restoration, but many also constrain how restoration is done. For example, river restoration projects in the United States nearly always require permits from the US Army Corps of Engineers due to the Clean Water Act of 1972. Restoration projects must consider potential negative environmental impacts, just like when developers want to build housing or roads. In coastal California, a small dam was removed to restore streamflow and allow passage of the *endangered* steelhead trout (*Oncorhynchus mykiss*), but the removal destroyed wetland habitat behind the dam that was habitat for the California red-legged frog (*Rana draytonii*), another species protected by the US Endangered Species

Act. This conflict required the creation of substitute habitat for the frog, greatly complicating the project. It is important for restoration *practitioners* to be well versed in the legal requirements for species and *ecosystem restoration* in their region, as well as the associated permits required to undertake restoration activities (chap. 3).

Here, I briefly discuss types of international agreements and national legislation relevant to restoration, but I do not provide lengthy discussions of individual regulations given that specific laws and their enforcement vary by country and municipality. I highlight some challenges to legislating restoration, note where approaches have been more or less effective, and suggest how legislation could be improved.

Background

The first international agreements (table 11.1) and national laws (table 11.2) that called for restoring damaged ecosystems were instituted approximately fifty years ago, and the number has increased dramatically since then. At the international level, these agreements are generally nonbinding and the definition of restoration is often vague (Telesetsky, Cliquet, and Akhtar-Khavari 2017), so they serve more as an aspirational framework than a set of enforceable *targets*. Those agreements, however, can provide guidance and incentives for national legislation. For example, the Convention on Biological Diversity, which stemmed from the Rio Summit in 1992, broadly called for the *rehabilitation* and restoration of degraded habitats. It was followed in 2010 by the Aichi Targets, which aim for the restoration of 15 percent of degraded habitats by 2020. The Aichi Targets have been incorporated into some national restoration plans (Telesetsky, Cliquet, and Akhtar-Khavari 2017). Mechanisms for enforcement at the international level are limited, so ultimately, laws are enforced at the regional (in the case of some European Union initiatives), national, state, or local level, although the extent of enforcement varies greatly. Laws are often established at the national level, but their specific requirements and enforcement vary by state or province. For example, the state of São Paulo in Brazil has defined clear vegetation recovery standards for evaluating farmers' compliance with the national Native Vegetation Protection Law, whereas the other twenty-six states in the country have not yet established standards (Chaves et al. 2015).

Most restoration legislation sets requirements for restoring specific habitat types, such as forest, wetlands, or arid systems (see tables 11.1 and 11.2). For example, the international Ramsar Convention and the US Clean Water Act protect wetlands because of their high habitat value and

Table 11.1. Examples of International Agreements That Call for Restoration[1]

Agreement	Year[2]	Requirement
Ramsar Convention	1971	Protect and restore wetlands of international importance
World Heritage Convention	1972	Protect and restore habitat within biosphere reserves
Convention on the Conservation of Migratory Species of Animals	1979	Restore habitat for migratory animals
United Nations Convention on Combatting Desertification	1994	Rehabilitate arid lands
Convention on Biological Diversity	1992	Rehabilitate and restore 15% of degraded habitats by 2020; more specific Aichi targets set later
Global Partnership on Forest and Landscape Restoration and Bonn Challenge	2011	Restore 150 million hectares of forest globally by 2020 and 350 million hectares by 2030

[1]These international agreements and more are described in detail in Telesetsky, Cliquet, and Akhtar-Khavari (2017).

[2]The year listed is the first year the agreement was established. In some cases, the agreements have been updated and expanded.

the many *ecosystem services* they provide to humans, such as water purification and flood control. Some legislation, such as the European Union Birds Directive and endangered species acts in various countries (see table 11.2), focuses on restoring habitat for and populations of specific *species of concern*. In other cases, regulations focus on establishing responsibility for restoration following certain environmentally degrading activities, such as mining or oil spills (see table 11.2).

A recurring problem with restoration legislation is that the definition and specific goals of restoration are often poorly articulated (Palmer and Ruhl 2015; Telesetsky, Cliquet, and Akhtar-Khavari 2017). Ecosystem restoration projects commonly focus more on restoring specific services, such as flood control and *carbon storage*, rather than restoring *community composition* (Palmer and Ruhl 2015). The term *restoration* has been used to refer to a range of activities, from *revegetating* eroded areas with minimal consideration for which species are planted to efforts that aim to restore the full suite of species that were present prior to *disturbance* (chap. 2). Hence, it is

Table 11.2. Examples[1] of Different Types of National or Regional[2] Laws Requiring Restoration

Law	Year	Country	Requirements related to restoration
Biosecurity Act	1993	New Zealand	Provides strict border controls aimed at reducing the entrance of invasive species into the country
Law for the Promotion of Nature Restoration	2002	Japan	Calls for a sound scientific underpinning for and stakeholder participation in restoration projects
Act on Conservation of Endangered Species of Wild Fauna and Flora	1992	Japan	Preserves endangered species of wild fauna and flora, as well as the natural environment those species depend on
Endangered Species Protection Act	1992	Australia	Promotes the recovery of species and ecological communities that are endangered or vulnerable
Native Vegetation Protection Law	2012	Brazil	Requires landowners to conserve or restore a certain percentage of forest on their land and maintain forest buffers along waterways
Water Framework Directive	2000	European Union	Mandates protection and enhancement of the status of aquatic ecosystems
German Federal Nature Conservation Act	1976	Germany	Requires polluter who causes an unavoidable impact on the landscape to minimize the impact and then compensate for remaining effects
National Policy for the Integral Management of Biodiversity and Its Ecosystem Services	2012	Colombia	Improves evaluation of environmental impacts, the recovery of environmental liabilities, and environmental offsets for biodiversity loss linked to environmentally licensable projects
Surface Mining Control and Reclamation Act	1977	United States	Requires developing a reclamation plan prior to coal surface mining and posting an environmental assurance bond that is returned when the reclamation plan has been completed successfully
Oil Pollution Control Act	2000	United States	Assigns liability for cleanup following oil spills

[1]Many countries have similar types of legislation.

[2]For the European Union.

critical to define clearly how terms are used in legislation and how success will be evaluated for specific projects.

Types of Legislation

Next, I describe different approaches to legislating restoration, which include preventing environmental damage, requiring the responsible party to restore a given site or undertake other actions to compensate for habitat destruction, and providing funding for restoration. Each is illustrated with examples.

Preventative Legislation

Although it does not fall strictly under the rubric of restoration, the most effective approach to conserving ecosystems is to protect them in the first place. This approach follows on the precautionary principle, which calls for not undertaking an action or carefully evaluating alternatives if there is potential harm to human health or the environment. For example, many countries regulate the import and transport of species that have the potential to become *invasive* because it is clear that efforts to prevent invasive species establishment are the most cost-effective (chap. 8). Given their many past problems with invasive species, Australia and New Zealand now have particularly strict invasive species prevention laws (chap. 8; Boonstra 2010; Eschen et al. 2015). Likewise, the Clean Water Act of 1972 in the United States regulates the discharges of pollutants into waterways "to maintain the chemical, physical, and biological integrity of the nation's waters." As a result, restoration projects that temporarily increase erosion during earth-moving activities must get discharge permits and take actions to minimize sediment inputs into waterways, such as conducting work during the dry season and using mulch or erosion control cloth.

Establishing Restoration Obligations

Many laws establish an obligation for the state or for specific *stakeholders*, such as private firms or landowners, to undertake restoration or other compensatory mitigation (discussed below) to reverse or minimize the effect of activities that have damaged ecosystems or harmed certain species. These laws aim to ensure that the responsibility for and costs of restoration are borne by the degrader. This practice follows the standard "polluter pays" principle in environmental law; in this case, it is a "degrader pays" approach (Telesetsky, Cliquet, and Akhtar-Khavari 2017). This obligation may be for past environmental impacts (e.g., leaking chemical storage tanks polluting rivers), anticipated actions (e.g., mining, construction), or potential future damages (e.g., oil spills).

Numerous laws set requirements for restoring specific habitat types. For example, Brazil's Native Vegetation Protection Law of 2012 (following earlier laws in 1934 and 1965) requires that private landowners maintain 20 to 80 percent (depending on the vegetation type and region) of their land in *native* vegetation and maintain native vegetation *buffer strips* along waterways (Atlantic Forest case study; Brancalion et al. 2016). Where the cover of native ecosystems has dropped below the requirements, landowners are required to restore the appropriate native forest, savanna, or grassland ecosystems within twenty years. Similarly, the European Union Habitats Directive of 1992 calls for restoration of habitat for flora and fauna with a focus on sites within the Natura 2000 network of protected areas in the EU (Telesetsky, Cliquet, and Akhtar-Khavari 2017).

Many countries have legislation requiring rehabilitation of habitat following mining, landfills, or other severely degrading activities. These efforts are commonly referred to as *reclamation*, because they usually focus on recovering ecosystem services, such as controlling erosion and improving water quality, rather than fully restoring the habitat. Occasionally, however, mine reclamation projects do aim to restore a diverse suite of native species. Mining companies are often required to outline a detailed reclamation plan prior to mining. For some types of mining in the United States, companies are required to deposit an *environmental assurance bond*, which is money that is held during mining and only released after the reclamation plan has been completed and deemed successful (Gerard 2000). This bond helps ensure that restoration is actually completed because part or all of the bond can be withheld if the reclamation is deemed as inadequate.

Compensatory Mitigation

Many countries have laws that require landowners to assess and offset (referred to as *compensatory mitigation* or *offsets*) the effects of projects that destroy specific habitats, species, or *ecosystem processes* (Maron et al. 2016; Cliquet 2017). These offsets often consist of actions to restore habitats or enhance the populations of species of concern, which is the focus of the examples here, although other types of offsets exist. *Mitigation* is a broader term that refers to a sequence of actions that aim to reduce the environmental impacts of a project. The mitigation sequence includes avoiding environmental damage to the greatest extent deemed possible, minimizing the impacts of the project, and compensating or offsetting impacts that cannot be avoided, either on the same or a different site. For example, under the US National Environmental Policy Act of 1970, all impacts are supposed to be avoided or minimized before habitat creation or restoration is considered as a compensation for impacts. The European Union Habitats Directive requires project developers to demonstrate the overriding public interest of a specific project before undertaking a project and associated compensation measures (Telesetsky, Cliquet, and Akhtar-Khavari 2017).

Most often, these laws prioritize restoring habitat or taking other actions to enhance populations of species of concern at the site where the degrading action takes place. An alternative approach is *mitigation banking* (also referred to as habitat banking), whereby the party responsible for causing environmental degradation pays a third party to undertake habitat protection, *management*, or restoration actions at another site (Marsh, Porter, and Slaveson 1996). Mitigation banking has the potential to result in larger habitat restoration projects that are implemented by groups with better expertise (Galatowitsch and Zedler 2014). In practice, however, mitigation banking often results in a net loss of habitat area when the protection of existing habitat is allowed as compensation for habitat destruction or when restoration projects fail (Brown and Lant 1999; National Research Council 2001). Moreover, mitigation banks may be located far from the site of environmental degradation, so the human, floral, and faunal populations directly impacted by the damage may not receive direct benefits (BenDor, Sholtes, and Doyle 2009). The international Ramsar Convention (see table 11.1) calls for a three-step approach of avoiding damage to wetlands, mitigating on-site, and then mitigating off-site as a last choice to conserve wetlands.

Many authors have questioned whether restoring habitat to offset losses that take place elsewhere provides the same habitat values and ecosystem processes and services as the original habitat (e.g., Maron et al. 2016; Schoukens and Cliquet 2016; May et al. 2017), and most studies to date support the conclusion that compensatory mitigation is not working. For example, Section 404 of the US Clean Water Act and a subsequent Memorandum of Understanding require mitigation for any wetlands that are degraded or destroyed. A review of these projects showed that, on average, mitigation efforts only compensated for approximately 20 percent of the ecosystem processes that were destroyed (National Research Council 2001) due to a mix of projects not being implemented, project failure, and a lack of *monitoring* of outcomes. Similarly, May, Hobbs, and Valentine (2017) found that few rehabilitation and restoration offset projects in Western Australia demonstrated the desired outcomes. Nonetheless, given the need to balance demands for human uses and efforts to conserve ecosystems and species, compensatory mitigation policies are widespread. Future compensatory mitigation policies should set clear objectives, monitor success, and include contingency plans for corrective actions if objectives are not achieved (chap. 4; May, Hobbs, and Valentine 2017) so that the projects actually provide the intended ecosystem services.

Providing Funding for Restoration

Many restoration laws at the national, state or province, and municipal levels commit public funds for habitat restoration and detail how the money will be spent (chap. 12). For example, the US Coastal Wetlands Planning, Protection and Restoration Act, passed initially in 1990 and renewed and expanded since then, requires that Louisiana prioritize and implement restoration projects on coastal wetlands and allocates federal and state funds for these efforts. If laws provide public funding for a certain habitat type in a region, then the relative costs and benefits of different projects should be assessed to most efficiently use limited restoration funds (chap. 12).

Challenges to Legislating Restoration

Designing appropriate, effective, and enforceable legislation is challenging. First, laws need to set specific *objectives* that must be met for compliance (chap. 4), but as discussed in chapter 5, ecosystems naturally show high variability in *ecosystem recovery* rates. Moreover, the choice of a *reference model* is complicated, and choosing an appropriate endpoint for future conditions is even more difficult given the rapidly changing climate (chap. 3). The challenge is to create laws that have clear *goals* and require a certain level of restoration while also recognizing natural variability and allowing flexibility to tailor restoration efforts to local conditions.

A second challenge is conflicting goals (chap. 3). As noted, endangered species legislation in many countries aims to restore populations of and habitats for species of concern, which may be at conflict with mandates to restore certain ecosystem types. For example, on the Colorado River, controlled releases of water from Glen Canyon Dam aim to restore sediment deposition patterns and habitat for native vegetation and fish species; these releases are being *adaptively managed* to minimize negative effects on the endangered Kanab ambersnail (*Oxyloma haydeni kanabensis*) and other species of concern that live below the dam and could be harmed by the high flows (Meretsky, Wegner, and Stevens 2000).

Third, the political time frame is much shorter than the time most ecosystems need to recover, which incentivizes short-term success. For example, success of coal surface mine reclamation in the southeastern United

States is generally evaluated after five years on the basis of the percent of plant ground cover and density of trees. The forest in this region, however, takes decades to recover. Planting aggressive, fast-growing grass species and one or a few tree species with high survival rates to meet requirements can inhibit long-term recovery of later *successional* vegetation (chap. 9; Holl 2002b). Ideally, legislation should call for setting specific project objectives that are evaluated at predefined intervals throughout the recovery process and that extend well into the future.

A fourth challenge to regulating restoration is land ownership, which strongly affects the success and longevity of restoration projects (Mansourian 2017). It is easier to hold private or public landowners accountable for their actions through legislation when ownership is clear than when it is not. In cases in which land ownership is communal or unclear, it is difficult to identify a party responsible for ecological damage and hold that party accountable for restoration.

Enforcement is another major concern with legislation. Although restoration and conservation laws are nearly always well intentioned, the key question is how well they are enforced. Systematic reviews of whether legislatively mandated restoration is effective show some successes, but more often, the evidence indicates that restoration efforts fall far short of targets. For example, the ambitious native vegetation restoration mandated by the various Brazilian Forest Codes has been weakly enforced in most of the country (Brancalion et al. 2016). In many cases, the extent of compliance with individual laws is not well known, given chronic problems with poor monitoring, reporting, and record keeping. For laws to be effective, the penalties for noncompliance should be large enough to provide an incentive to the responsible party to comply, and those penalties need to be enforced.

Improving Future Restoration Legislation

Clearly, there is a great deal of room for improvement in how restoration laws and policies are designed, as well as how they are implemented and enforced. Evaluation of a wide suite of legislation suggests some key steps to improving the outcomes in the future. First, as noted previously, it is critical to define clearly how specific terms are used and what goals and measurable objectives the legislation aims to achieve. Second, to increase success and minimize unwanted side effects, legislation should incorporate the best available information based on both scientific studies and past restoration projects. This aim can be achieved by consulting with both scientists and restoration experts when developing the legislation and requiring peer review of project plans as part of the legislation. Moreover, natural resource managers should be required to summarize and publicly share the outcomes of their restoration efforts, both to ensure accountability and to improve future restoration efforts.

Legislation should require monitoring to evaluate whether objectives have been achieved and incentives to ensure that corrective actions are taken if those are not met, in order to complete the adaptive management cycle (chap. 4; May, Hobbs, and Valentine 2017). Ideally, monitoring should be done by an independent third party, who is paid by the municipality rather than by the group responsible for the restoration. This increases the likelihood of rigorous evaluation of whether the objectives have been met, since the person monitoring the project does not have a vested interest in whether the project is judged as successful.

Finally, restored sites should be protected in perpetuity. Many laws and international agreements set targets for how much land should be restored, but do not ensure whether those ecosystems remain over the longer term. Reid et al. (2019) found that 50 percent of recovering forest area in Costa Rica was recleared within twenty years, highlighting the importance of long-term protection and monitoring of restored ecosystems to ensure that they deliver the conservation values and ecosystem services, such as water supply and carbon storage, that motivate their restoration.

Recommended Reading

Cliquet, An. 2017. "International law and policy on restoration." In *Routledge Handbook of Ecological and Environmental Restoration*, edited by S. K. Allison and S. D. Murphy, 381–400. London: Routledge.
   Summarizes international laws and policies related to ecological restoration.

Mansourian, Stephanie. 2017. "Governance and restoration." In *Routledge Handbook of Ecological and Environmental Restoration*, edited by S. K. Allison and S. D. Murphy, 401–13. London: Routledge.
   Reviews governance issues related to ecological restoration.

Telesetsky, Anastasia, An Cliquet, and Afshin Akhtar-Khavari. 2017. *Ecological Restoration in International Environmental Law*. London: Routledge.
   Provides a detailed discussion of international environmental laws and agreements related to ecological restoration.

12

Paying for Restoration

Although there are many biophysical and social barriers to *ecological restoration* (chap. 2), insufficient funding is a major issue. Restoration is often expensive. Restoring the Kissimmee River (case study) cost nearly $5 million per kilometer of river. Bayraktarov et al. (2016) reported that the median cost for marine coastal restoration was approximately $80,000 per hectare (in 2010 dollars), although the cost of individual projects varied greatly depending on the country and *ecosystem* type (e.g., mangrove vs. coral reef). Many restoration projects are considerably cheaper, particularly when there is less initial engineering and site preparation and more reliance on volunteer labor, but even for less expensive restoration methods, the costs mount quickly given the large scale of restoration proposed in some international agreements (chap. 11). Pistorius and Freiberg (2014) estimate that complying with the Aichi target to restore 15 percent of degraded habitats worldwide would cost between $45 billion and $75 billion even at the low cost of $500 to $1,500 per hectare. Nonetheless, in most cases, restored ecosystems provide benefits to humans that far outweigh the costs of restoration (De Groot et al. 2013).

Restoration incurs a host of costs, including salaries for those involved in planning, *stakeholder* coordination, and project management; equipment, materials, transportation, and staffing for site preparation and construction; acquisition and introduction of plants or animals; and site *maintenance* and *monitoring*. It is often necessary to purchase land or to compensate landowners for the revenue lost from using their land for other income-generating activities. Monitoring costs should be included in project budgets, along with funding for contingencies when restoration does not go according to plan, although contingency funding is rarely available (Bayraktarov et al. 2016; Iftekhar et al. 2017).

In this chapter, I first discuss the rationale for investing substantial sums of money in restoration. I then turn to potential sources of funding and close by discussing several strategies for most effectively raising and allocating funding for restoration.

Benefits of Investing in Restoration

Although restoration projects are often expensive, they can provide enormous benefits to humans. Many large-scale restoration efforts, particularly those with public financing, are justified based on the *ecosystem services* they provide to society, such as flood control, *carbon storage*, and coastal hazard risk reduction (Aronson, Milton, and Blignaut 2007; De Groot et al. 2013; Ferrario et al. 2014). Likewise, funding for *invasive species* control has focused on removing species that have clear negative economic consequences, such as plants and insects that reduce agricultural and rangeland *productivity* or aquatic organisms that clog power plant water intake systems.

Increasingly, restoration funding is justified because restored ecosystems provide certain ecosystem services to humans more cheaply than engineering alternatives. For example, Ferrario et al. (2014) report that the median cost of installing breakwaters to reduce coastal hazards in tropical regions is $21,000 per meter, whereas restoring coral reefs to provide a similar service costs an order of magnitude less ($1,300 per meter). Tropical forest restoration has been repeatedly suggested as one of the most cost-effective approaches to sequester carbon, although funding available through this mechanism has been much smaller than anticipated (Brancalion et al. 2017). Restoring forests provides numerous additional services to humans besides carbon storage, such as improving water quality and providing timber and nontimber products (Ding et al. 2017).

In some cases, the value of these ecosystem services translates directly into funding for restoration through *payments for ecosystem services* offered to farmers or landowners as an incentive to *manage* their land to provide specific ecosystem services. Costa Rica had one of the earliest programs to pay landowners to conserve and restore forest (Pagiola 2008); funding for this program has come from domestic taxes on fossil fuels and water, as well as international funding from the World Bank and from individual countries to promote *biodiversity* conservation, water quality, and carbon storage. In recent years, payments for ecosystem services programs that support restoration have been instituted in numerous countries worldwide. Restoration is also justified by the employment, training, and economic opportunities it creates (van Wilgen and Wannenburgh 2016), especially in rural communities far from urban centers. Community-based programs, such as mangrove restoration in Sri Lanka (Wickramasinghe 2017), purchase seedlings from local nurseries and employ community members to plant and maintain seedlings (Asian Mangrove case study). BenDor et al. (2015) estimate that for every restoration job created there is a job multiplier of an additional 1.5 to 3.8 jobs created in the economy.

Although it is easier to make the case for large expenditures on restoration when the economic benefits can be easily quantified, many other benefits of restoring ecosystems are harder to measure. They include species conservation and cultural services, such as the emotional values associated with the aesthetics, community pride, and sense of stewardship that come from experiencing and protecting natural spaces. In addition, the cultural uses of an area can be improved through restoration, as is often true for hunting and fishing, recreation, and the maintenance of sacred sites for indigenous communities. These uses all can be important motivators for investing time and money in restoration (chap. 1), but are difficult to quantify monetarily. One commonly used method for valuing restoration's intangible benefits is *contingent valuation*, an approach in which people are surveyed about their willingness to pay to conserve or restore an ecosystem or specific species (Holl and Howarth 2000; Iftekhar et al. 2017). Nonetheless, it remains difficult to translate the full range of values that humans place on intact ecosystems into specific monetary values to invest in restoring ecosystems, and even if these benefits are quantified, it does not guarantee that funding will be forthcoming.

Who Pays for Restoration?

Restoration projects are funded by a variety of sources, depending on whether a party is held responsible for prior *degradation*, the *goals* and scale of the restoration project, and who benefits or is willing to pay for the restoration (fig. 12.1; Holl and Howarth 2000). A brief description of common funding sources and examples of these sources are described next.

Funding by Responsible Party

If an individual, group of people, corporation, or government agency is clearly responsible for degradation of a given ecosystem, then that party should pay for restoration (see fig. 12.1). Many laws are based on the

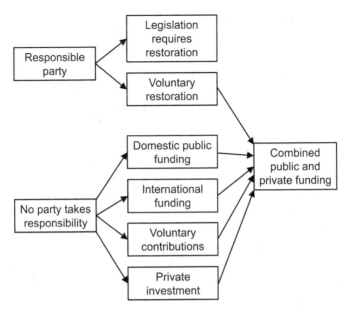

Figure 12.1. Funding sources for restoration. Examples of each type of funding are provided in the text. Modified from Holl and Howarth 2000.

principle of "the degrader pays" (chap. 11), such as laws protecting specific threatened or *endangered species* or habitats or regulating types of activities (e.g., mining). These laws hold the responsible party accountable for the full costs of environmental damage and restoration. For example, in the Younger Lagoon case study, the landowner is required to pay the restoration costs to comply with several US laws that protect threatened habitats and species (see table 3.1; chap. 11). Ideally, this accountability should be secured through *environmental assurance bonding* before the environmental damage takes place and should include a clear system of penalties for noncompliance (chap. 11; Costanza and Cornwell 1992; Gerard 2000). The financial liability or incentive to successfully complete the project should be sufficiently large to ensure a reasonable attempt to restore the ecosystem.

Although most restoration undertaken by the party responsible for the degradation is done because of legislative mandates, some parties undertake restoration voluntarily because of a desire to improve land or water resources, to enhance the public image of a company, or to fulfill a sense of ethical responsibility (Telesetsky 2017). For example, CEMEX, an international cement company, signed onto the Cancun Business and Biodiversity Pledge committing to work to conserve biodiversity (Telesetsky 2017). In addition to undertaking quarry *rehabilitation* projects, CEMEX has managed a private 140,000-hectare reserve on the US-Mexico border to protect and restore habitat for threatened species such as the desert bighorn sheep (*Ovis canadensis nelson*). Many smaller landowners remove invasive species and replant *native* vegetation on their land to restore habitat, water quality, and aesthetic values. Often, landowners or organizations receive financial incentives from the government for voluntary restoration (discussed below).

Although the responsible party should pay for restoration, this principle can be difficult to apply in practice. In many cases, the degradation occurred so long ago that it is difficult to identify the responsible party. Further, environmental damage often is caused by cumulative effects of many small businesses or individuals, making it impossible to declare one party at fault. For example, it is rarely possible to identify a single party who is responsible for the introduction of an invasive species that subsequently requires substantial cost and effort to remove. Likewise, in the case of *nonpoint source pollutants*, such as animal feces, motor oil, and lawn fertilizers, distributed discharge by many people causes water quality problems in urban areas, but no one person is at fault. Other sources of funding are needed in cases where it is difficult to assign responsibility for environmental damage or when the party that caused the degradation is not held responsible for restoration (see fig. 12.1).

Domestic Public Funding

Taxes are commonly used to pay for restoration. In some cases, specific raw and manufactured materials or environmentally damaging actions are taxed to reflect their environmental costs. This approach serves to internalize some of the costs of degrading activities and to provide a price that more accurately reflects the true environmental cost. For example, under the National Coastal Wetlands Conservation Grant Program of the US Fish and Wildlife Service, excise taxes from the sale of fishing equipment, motorboats, and fuel for small engines help pay for the acquisition and restoration of coastal marshes. More recently, some countries and municipalities are adopting policies to place a price on carbon emissions that could be used to fund restoration. For instance, restored *riparian* forests in California store enough carbon in the first two decades after restoration that payments through California's carbon emissions policies are sufficient to pay for the restoration, if the land is already in conservation ownership (Matzek, Puleston, and Gunn 2015).

Many large-scale restoration projects are funded by general taxes that do not target the specific cause of destruction. For example, the Kissimmee River Restoration Project (Kissimmee River case study) is supported primarily by taxpayer funds at both the national and state levels; in this case, ironically, the US Army Corps of Engineers, a government agency funded by taxpayers, caused the original damage. Similarly, many invasive control efforts and other small-scale restoration projects are supported by the budgets of national and municipal natural resource management agencies.

Government support often takes the form of matching funds or tax incentives to private landowners who voluntarily undertake restoration actions on their property. For example, agri-environment programs in both the United States and Europe pay farmers to remove wetlands and other sensitive habitats from crop production, to restore these habitats in some cases (Galatowitsch and Zedler 2014), and to adopt farming practices that restore *landscape connectivity* within the agricultural landscape (Rey Benayas and Bullock 2012). Between 1999 and 2015, the Chinese government paid farmers cash incentives to reestablish forest, shrub, or grassland ecosystems on 278,000 square kilometers of degraded agricultural lands (an area the size of Ecuador) through the "Grain-for-Green" program, the largest *reforestation* program in the world (Hua et al. 2016). The vast majority of the reforested area, however, was planted with monocultures or a few species with relatively low biodiversity value.

International Government Funding

At the global scale, many projects are funded by grants or low-cost loans from intergovernmental organizations, such as the Global Environment Facility of the World Bank, or by international aid funds from the government of a high-income country to a lower-income country. These projects are typically targeted at improving human *livelihoods* and environmental quality in lower-income countries, as well as storing carbon and improving water supply and quality (Ding et al. 2017). For example, the World Bank provided $9.5 million to Rwanda between 2014 and 2019 to restore riparian forests to improve water supply and climate resilience and to promote agricultural practices that enhance farmers' livelihoods and environmental sustainability (World Bank n.d.). Similarly, the governments of western European countries, Japan, and the United States have contributed $135 million to the second phase of an initiative to restore 2 million hectares of peatlands in Indonesia (Hansson and Dargusch 2017), although it is only a small fraction of the estimated $4.6 billion needed to restore the entire peatland system.

Voluntary Contributions of Funding and Labor

Many restoration projects, both domestically and internationally, are funded by the philanthropic contributions of nonprofit organizations or wealthy individual private donors. Large international conservation organizations, such as Conservation International, The Nature Conservancy, and the World Wildlife Fund, provide extensive funding for restoration projects around the globe. Bremer et al. (2016) reported that nonprofit groups provided more than $10 million between 2014 and 2016 to conserve and restore *watersheds* in Latin America. These organizations receive most of their funding from private donors who often receive tax incentives for charitable donations. Other examples include the Tompkins Family Foundation, which has supported the conservation and restoration of several parks in Chile and Argentina (Tompkins Conservation n.d.), and media mogul Ted Turner, who is working to *reintroduce* endangered species and restore habitat on the 800,000 hectares of land he owns in the United States (Turner Endangered Species Fund n.d.).

Voluntary monetary support is only one type of contribution that private individuals can make toward restoration; many smaller restoration projects rely partially or entirely on volunteer labor (Berger 1985). In New Zealand, more than six hundred community groups are involved in restoration that relies heavily on volunteers; these projects primarily aim to restore native flora and fauna and provide environmental education (Peters, Hamilton, and Eames 2015). Community participation in restoration projects has the added benefits of educating volunteers, developing their sense of stewardship, and often developing an advocacy group for the restoration efforts.

Private Investment

There has been growing discussion of the need to increase private restoration funding from for-profit businesses and investors in order to restore at the large scale of recent international commitments (Brancalion et al. 2017; Ding et al. 2017). Securing private investors has proven to be challenging because many of the ecological and social benefits of restoration do not have a clear market value to generate profits for the investor, most restoration projects are too small, the timeline for return on investments is too long, and restoration is considered too risky to attract private investors (Ding et al. 2017). Nonetheless, there are some recent examples of private investment in restoration, particularly for carbon storage. Moreover, some businesses are funding restoration as part of their corporate social

responsibility programs (Telesetsky, Cliquet, and Akhtar-Khavari 2017).
WeForest, a nonprofit organization, has coordinated more than 140 private companies from 24 countries to invest in projects to restore tree cover
with ecological and social benefits (Gutierrez and Keijzer 2015). Businesses
often support restoration projects by donating to nonprofit restoration organizations so as to receive both tax incentives and to be able to market
their products as environmentally friendly. For example, many Brazilian
companies, ranging from banks to beer companies, have supported the
Atlantic forest restoration and market themselves as environmentally con-
scious (Atlantic Forest case study).

Combined Public and Private Funding

Many large-scale restoration projects are supported by both public and private funding. For example, restoration of mangroves in Asia (Asian Mangrove case study) and restoration of the Brazilian Atlantic forest (Atlantic
Forest case study) have been funded by international sources, both governments and nonprofit organizations, along with domestic government
funding and private donors. The Corporate Wetland Restoration Program
in New Jersey provides funding for wetland restoration through donations
of both volunteer time and materials from corporate partners, funding from the private sector, and funding from federal and state budgets
(NJCWRP n.d.). Private contributions may consist of land that is donated
or sold at a low cost and subsequently restored with funding from nonprofit organizations or public agencies.

Strategies for Increasing Restoration Funding

The need for restoration continues to grow, but available funding remains far below what is required. There are, however, several strategies
that would expand the amount of funding available and allocate it more
efficiently.

Internalizing Externalities and Redirecting Subsidies

Two key actions to better account for the value of ecosystem restoration
include internalizing the costs of land degradation and removing the subsidies that support ecosystem degradation. Ecosystem degradation has
many negative *externalities* (i.e., consequences of anthropogenic activities
that are unaccounted for in the cost of a good or service), such as the
contamination of soil and water, which are rarely charged to the responsible party. In contrast, restoration often provides positive externalities that
are not quantified. Moreover, many countries provide billions of dollars

in *subsidies* each year, which lead to substantial deforestation and unsustainable agricultural practices and mineral extraction (Ding et al. 2017).
Reducing subsidies that support environmentally degrading practices, taxing products to reflect their true environmental cost, and redirecting funds to ecological restoration are challenging but critical actions that would help countries move toward meeting their large-scale restoration commitments. Making these changes would require governments to adopt tax law revisions and new regulations.

Allocating Restoration Funds Strategically

For large-scale restoration programs, particularly those that are publicly funded, it is important to evaluate the relative costs and benefits of individual projects in a region. One factor to consider is the lost *opportunity costs* of using the land for income-generating purposes, which varies greatly across a region, making restoration cheaper and more feasible in sites that are not highly valued for other human uses, such as agriculture or housing (Latawiec et al. 2015). Likewise, the cost of restoring a given ecosystem type varies greatly from site to site depending on the level of degradation, proximity to *propagules* to enable *natural regeneration*, and whether costly *active restoration* methods are needed (Holl and Aide 2011). Bayraktarov et al. (2016) found that project location and restoration methods more strongly affected marine coastal restoration success than did the total cost, suggesting that choosing restoration sites and methods carefully would result in more effective allocation of funds. Similarly, Strassburg et al. (2019) reported that using a prioritization scheme could reduce restoration costs across the Brazilian Atlantic forest biome by 57 percent while at the same time improving gains in carbon storage and biodiversity conservation. Although such prioritization exercises help evaluate the most judicious expenditure of limited funds across a region, they must be balanced with localized reasons for restoring a given site (e.g., restoration that is legislatively mandated for a specific landowner, presence of a specific *species of concern*, volunteer-led restoration efforts at a given location).

Detailed accounting of costs combined with comparisons of the success of different restoration methods can help reduce costs within individual projects. Kimball et al. (2015) compared the cost-effectiveness of numerous restoration methods to restore coastal grassland and sage scrub in southern California. They found that spending money to reintroduce and maintain vegetation increased native plant cover more than paying a similar amount for site preparation. They used their results to develop a decision tree to help *natural resource managers* select the most cost-effective methods given their site conditions and restoration goals.

Exploring Potential Income Sources from Restoration

Going forward, we must think innovatively about ways to generate revenue from restoration. For example, scientists and land managers have been exploring models to offset short-term tree-planting costs to restore tropical forests by including either small-scale agricultural plantings or valuable timber trees among their native species plantings (Atlantic Forest case study; Vieira, Holl, and Peneireiro 2009); crops can be harvested for a decade or so as a transitional phase in the restoration process, whereas timber and nontimber forest products can be selectively harvested over a range of time intervals. Brancalion, Viani, Strassburg, et al. (2012) outline several sources of income from tropical forest restoration efforts through a mix of payments for ecosystem services, nontimber forest products, and selective logging of timber.

Ensuring Long-Term Funding

The most difficult issue in assigning restoration responsibility and costs is uncertainty over time. It is difficult to estimate the time frame of ecosystem *recovery* given unpredictable events such as extreme weather conditions, pest outbreaks, and long-distance seed dispersal, coupled with our limited scientific knowledge. Questions to address prior to starting any restoration project include: For how much money and for how long should a landowner be held responsible? Who will pay for additional costs in the likely situation that restoration efforts do not go according to plan? In most cases, the parties who undertake restoration are held responsible for only a few years until the implementation stage of a restoration project is completed, well before when *objectives* have been met.

For restoration efforts to succeed far into the future, we must develop mechanisms that recognize uncertainty and acknowledge that restoration is a long-term endeavor. Even some well-planned restoration projects with realistic cost estimates and sufficient funding will fail due to extreme climatic conditions, such as a hurricane, or other unexpected occurrences, such as a pest outbreak. In most past cases where restoration efforts have not met project objectives, either no corrective actions have been taken or unanticipated and long-term costs have been covered by public funding. Going forward, it should be made explicit in restoration plans which party is responsible for cost overruns, and some funding should be allocated to implement corrective actions through the *adaptive management* cycle (chap. 4). One approach that has been used to provide long-term funding, particularly in cases when restoration is done as *compensatory mitigation*, is to require that developers, homeowners, or other responsible parties pay into an endowment fund at the time the project starts and then use the payout for ongoing project maintenance and monitoring. Another interesting example of long-term funding is from Quintana Roo, Mexico, where taxes on the coastal tourism industry are being used to purchase an insurance policy that will pay to restore coral reefs following hurricane damage (Nature Conservancy 2018). Resolving the mismatch between the short political and budgetary timelines and the much longer time needed for ecosystems to recover will be an ongoing challenge.

Recommended Reading

Aronson, James, Suzanne J. Milton, and James N. Blignaut. 2007. *Restoring Natural Capital*. Washington, DC: Island Press.

> Discusses the general benefits of restoring ecosystems and presents many detailed case studies.

Ding, Helen, Sofia Faruqi, Andrew Wu, Juan C. Altamirano, Andrés Anchondo Ortega, Michael Verdone, René Zamora Cristales, et al. 2017. *Roots of Prosperity*. Washington, DC: World Resources Institute.

> Provides a detailed overview with many examples of the costs and benefits of forest restoration, obstacles to investing in forest restoration, and ways to increase investment.

Iftekhar, M. S., Maksem Polyakov, Dean Ansell, Fiona Gibson, and Geoffrey M. Kay. 2017. "How economics can further the success of ecological restoration." *Conservation Biology* 31:261–68.

> Discusses issues related to assessing social and economic benefits of restoration, estimating overall costs, prioritizing and selecting projects, and securing long-term funding.

Glossary

The definitions here are largely consistent with those in Gann et al. (2019).

Abiotic Nonliving materials and conditions within an ecosystem, including rock, water, the atmosphere, weather and climate, *topography*, *hydrology*, fire, and salinity regimes.

Active restoration (also Reconstruction) A restoration approach in which there is extensive human intervention to influence the rate and *trajectory* of recovery and the arrival of the biota is largely or entirely dependent on human agency.

Adaptive management An ongoing process for improving management practices by applying knowledge learned through assessment of previously employed practices to improve current and future projects. The practice of revisiting management decisions and revising them in the light of new information.

Alternative states Alternative ecosystem states and environmental conditions that may persist at a particular spatial extent and temporal scale. The probability and number of alternative states associated with any given ecosystem varies.

Assembly rules A set of principles or theories that describe the development of biological *communities* from a larger regional pool of potential contributor species. Theories of ecosystem assembly have changed considerably over time, and biologists now understand that environmental filters, *biotic* interactions, and stochastic processes each play a role.

Assisted regeneration An approach to restoration that focuses on actively triggering any *natural regeneration* capacity of biota remaining on site or nearby, as distinct from reintroducing the biota to the site or leaving a site to regenerate naturally. Although this approach is typically applied to sites of low to intermediate degradation, some highly degraded sites have proven capable of assisted regeneration given appropriate treatment and sufficient time frames. Interventions include removal of pest organisms, managing ecological disturbance regimes, and installation of resources to prompt colonization.

Ballast water   Water carried in tanks of ships to improve stability and balance. This water is taken up or discharged when cargo is unloaded or loaded or when a ship needs extra stability in foul weather.

Baseline inventory   A description of current *biotic* and *abiotic* elements of a site prior to *ecological restoration*, including its compositional, structural, and functional attributes. The inventory is implemented at the commencement of the restoration planning stage, along with the development of a *reference model*, to inform planning, including restoration *goals*, measurable *objectives*, and treatment prescriptions.

Best management practices   A practice or combination of practices, that are effective and practical means of preventing or reducing the amount of pollution generated by nonpoint sources to a level compatible with water quality goals. Sometimes used more broadly to refer to sets of practices considered to represent the current, highest standard practices in any area of management or restoration.

Biodiversity   The variability among living organisms from all sources, including terrestrial, marine, and other aquatic ecosystems and the ecological complexes of which they are part; includes diversity within species, between species, and of ecosystems.

Biological control   The introduction of an herbivore, predator, or pathogen to control the population of an unwanted target organism. Substantial testing is required to ensure host specificity before the biological control agent is released.

Biological soil crusts   Communities of fungi, lichens, cyanobacteria, bryophytes, and algae that form a crust on the soil surface in some arid systems. Soil crusts help with primary production, nitrogen fixation, and soil stabilization.

Bioremediation   The use of living organisms to treat toxic wastes or remediate contaminated soil, water, or air.

Biotic, biota   The living components of an ecosystem, including the living animals and plants, fungi, bacteria, and other forms of life, from microscopic to large.

Buffer strips   Narrow strips of vegetation adjacent to wetlands or rivers that serve to filter sediments and pollutants from nearby land uses and provide flood protection services and riparian habitat.

Captive breeding   Maintaining populations of animals in captivity, such as at zoos or within reserves, for breeding purposes to increase their populations and to provide stock for reintroduction projects.

Carbon storage   The capture and long-term storage of atmospheric carbon dioxide, typically in biomass accumulation by way of photosynthesis, vegetation growth, and soil organic matter buildup. May occur naturally or be the result of actions to reduce the rate of climate change.

Channelization   Straightening of rivers to create more navigable waterways and, when accompanied by channel deepening and levees, to provide flood control.

Check dam (also Weir)   A small, low dam or obstruction across a small stream meant to slow water flow and increase water depth.

Community composition   The array of organisms within an ecosystem.

Compensatory mitigation (also Offsets)   Measures required by government agencies, or international agreements, to obtain permission for development projects that cause unavoidable environmental harm. These measures aim to compensate for damage or destruction done to a site or an ecosystem by expanding existing protected areas or undertaking ecological rehabilitation, ecological restoration, or habitat creation, often in a different area from where the damage occurs.

Compete/Competition/Competitors   A mutually negative interaction between two organisms that share the same resources.

Contingent valuation   A method of evaluating the value a person places on a good or service, such as the restoration of an ecosystem or species, by surveying individuals on their willingness to pay for this good or service. This approach contrasts with only using values that can be quantified directly in marketplaces.

Culvert   A tunnel or channel that carries a stream beneath a pathway, road, or other obstruction that would otherwise interrupt water flow.

Cycling (ecological)   The transfer between parts of an ecosystem of resources such as water, nutrients, and other elements that are fundamental to all other ecosystem processes.

Degradation (of an ecosystem)   A level of deleterious human impact to ecosystems that results in the loss of *biodiversity* and simplification or disruption in their composition, structure, and functionality and generally leads to a reduction in the flow of ecosystem goods and services.

Disturbance regime   The pattern, frequency, and timing of disturbance events that are characteristic of an ecosystem over a period of time.

Disturbances   Natural or human-mediated events or activities that change the structure, species composition, and/or function of an ecosystem. Disturbances may be beneficial or deleterious, depending on their type, intensity, scale, and frequency with respect to adaptations of particular organisms.

Drill seeding   The use of a drill to create shallow depressions into which the seed is deposited. Drill seeding can be followed by tamping to cover the seeds with a thin layer of soil.

Ecological corridors   Strips of habitat that link and allow the movement of plant and animal species between isolated habitat areas.

Ecological restoration (also Ecosystem restoration)   The process of assisting the *recovery* of an ecosystem that has been degraded, damaged, or destroyed.

Ecological trap   A *habitat* area that attracts organisms away from their source habitat but in which they struggle to survive and reproduce.

Ecosystem   A small- or large-scale assemblage of *biotic* and *abiotic* components in water bodies and on land in which the components interact to form complex food webs, nutrient cycles, and energy flows.

Ecosystem function   *See* Ecosystem process.

Ecosystem maintenance   Ongoing activities, applied after full or partial recovery, intended to counteract processes of ecological degradation and to sustain the attributes of an ecosystem. More ongoing maintenance is likely to be required at restored sites where higher levels of threats continue compared to sites where threats have been controlled.

Ecosystem process (also Ecosystem function)   An intrinsic ecosystem characteristic whereby an ecosystem maintains its integrity. Ecosystem processes include decomposition, production, nutrient cycling, and emergent properties resulting from species interactions such as competition, seed dispersal carried out by animals, and mutualistic relationships. Some ecosystem processes can deliver *ecosystem services* and goods to humans.

Ecosystem services   The direct and indirect contributions of ecosystems to human well-being. They include the production and maintenance of clean soil, water, and air; the moderation of climate and disease; nutrient cycling and pollination; the provisioning of a range of goods useful to humans; and potential for the satisfaction of aesthetic, recreation, and other human values.

Ecosystem structure   The physical organization of an ecological system, including density, stratification, and distribution of organisms (their populations, habitat size, and complexity); canopy structure; pattern of habitat patches; and abiotic elements.

Ecotype   A genetically and generally physiologically or morphologically distinct subset of populations within a species that provides its organisms with an adaptive advantage to a location and its particular environmental conditions.

Endangered species   Plant and animal species considered to be at high risk of *extinction*. Many individual countries and local governments have adopted their own legal definitions to list and protect such species.

Environmental assurance bond   Money paid by an individual or organization that plans to cause environmental damage, the payment of which is meant to ensure that the responsible party rectifies their actions postdegradation. This money is held in an account until the responsible party has successfully completed the agreed-upon reclamation or restoration plan, at which time it is returned. If the reclamation plan is not completed, then the money is not returned.

Environmental gradient   A gradual change in abiotic conditions through space or time. Environmental gradients can be related to altitude, temperature, soil or water depth, ocean proximity, salinity, soil moisture, or other abiotic factors.

Environmental stochasticity   The random variation in natural processes that occur in the environment, such as variation in precipitation, temperature, water flow, or natural disturbances such as fire or flooding.

Eradication   Removal of every individual and propagule of an invasive species.

Externality   A side effect or consequence of an industrial or commercial activity that affects other parties but is not reflected in the cost of the goods or services produced.

Extirpated   When a species no longer exists within a certain geographical location, but still exists elsewhere.

Eutrophication   An excess of nutrients in an ecosystem, often caused by anthropogenic enrichment of waters, most commonly phosphorus, beyond their natural or historic levels. It is particularly evident in shallow lakes and estuaries.

Exotic Species   *See* Nonnative species.

Extinction   The termination of any lineage of organisms, from subspecies to species and higher taxonomic categories from genera to phyla.

Facilitate/Facilitation   A species interaction in which one species benefits and the other neither benefits nor is harmed.

Fish ladder   A series of pools built like steps to enable fish to bypass a dam.

Fitness   The net reproductive output of an organism.

Floodplain   The region of low-lying land adjacent to a river, typically composed of high nutrient sediments and subject to regular flooding in the absence of human intervention.

Focal species   Species chosen as a focus for restoration or conservation actions.

Forest and landscape restoration   A planned process that aims to regain ecological functionality and enhance human well-being in deforested or degraded landscapes.

Framework species method   An ecological restoration strategy that involves reintroducing the minimum number of species required to reinstate ecosystem structure and processes and to enable recolonization by additional species from adjacent areas. In forest ecosystems, it often combines the planting of several species that attract fauna and are from different stages of succession.

Gene flow   Exchange of genetic material between individual organisms that maintains the genetic diversity of a species' population or *metapopulation*. Gene flow can be limited by dispersal vectors and by topographic barriers such as mountains and rivers. In fragmented landscapes, it can be limited by the separation of remnant *habitat*.

Genetic diversity   The measure of genetic characteristics (or genotypes) and their relative abundances within a population of organisms.

Goals (also Target)   Ecological and social outcomes sought at the end of a restoration project.

Habitat   The natural environment (including specific abiotic and biotic conditions) in which a species grows or lives.

Hardening (of plants)   The process of preparing greenhouse-grown plants for the stress of the natural environment by subjecting them to field-realistic levels of sunlight, moisture, and temperature.

Hard release   Releasing wild or captive-bred animal populations into the wild with no prior training for or exposure to field conditions.

Hedgerow   Lines of closely space shrubs or trees at the edge of agricultural fields that often serve as property boundaries, facilitate the movement of some animals, and serve as *buffers strips* to filter water.

Hydrologic regime   The timing and magnitude of water-flow patterns in aquatic systems.

Hydrology   The scientific study of waters above and below the land surfaces of Earth; their occurrence, circulation, and distribution, both in time and space; their biological, chemical, and physical properties; and their reaction with their environment, including their relation to living beings.

Hydroperiod   The depth, duration, frequency, and seasonality of water levels in wetlands.

Hydroseeding The use of water or other liquids to place seed by mixing the seeds in the liquid and spraying it over the desired sowing location.

Inbreeding depression The process by which deleterious genes accumulate in the offspring of organisms with the same genetic makeup, resulting in reduced likelihood of survival and breeding.

Inundation The flooding of an area of land, either temporarily or permanently.

Integrated pest management An ecosystem-based strategy that focuses on long-term prevention of *invasive species* and their damage through a combination of techniques such as physical removal, biological control, and habitat manipulation. Chemical control methods are used only after monitoring indicates that they are needed according to established guidelines and are selected and applied in a manner that minimizes risks to human health, beneficial and nontarget organisms, and the environment.

Invasive species Species that spread rapidly and have the capacity to dominate available *habitats* to the detriment of *native species, ecosystem processes*, and *ecosystem services*. Invasive species are primarily *nonnative*, but the term sometimes is used to refer to aggressive native species for which the population is growing rapidly due to anthropogenic impacts.

Landform A natural physical feature of Earth's surface.

Landscape connectivity The degree to which the configuration of features in a landscape *facilitate* or impede the movement of organisms between patches of *habitat*.

Levee The reinforced edges along a stream or river designed to contain water during high flow events. Levees can be made of natural or human-made materials.

Livelihood The capabilities, assets (including both material and social resources), and activities required for a means of living.

Local ecological knowledge Knowledge, practices, and beliefs regarding ecological relationships that are gained through extensive personal observation of and interaction with local *ecosystems* and shared among local resource users.

Management (of an ecosystem) A broad categorization that can include maintenance and repair of ecosystems, including *restoration*.

Metapopulation A set of partially isolated subpopulations of a given species. Long-term survival of the species depends on a shifting balance between local *extinctions* and recolonizations.

Microcatchment A small depression in a land surface designed to concentrate surface runoff of rainfall, plus nutrients, detritus, and seeds. Often used in management of arid and semiarid systems.

Microsite A small part of an *ecosystem* with a unique set of features and conditions that differs markedly from its immediate surroundings, usually at a small (meter to centimeter) scale.

Mitigation A series of actions taken to minimize the environmental damage of a development or danger to a *species of concern*. The potential steps involved in mitigation include avoiding project alternatives that would be particularly damaging, modifying the project to minimize negative impacts to the degree possible, and compensating for or offsetting impacts that cannot be avoided through *compensatory mitigation*. In the context of climate change, refers to reducing emissions of and stabilizing the levels of heat-trapping greenhouse gases in the atmosphere.

Mitigation banking (also Habitat banking) The restoration or protection of *ecosystems* or *habitats* by a third party that may be sold to a group to compensate for future damage to an ecosystem or loss of habitat for one or more *species of concern*.

Monitoring The systematic and orderly gathering of data over a period of time so as to evaluate whether specific project *objectives* are achieved.

Mutualism A mutually beneficial interaction between species, such as seed dispersal by fauna and *mycorrhizae*.

Mycorrhizae A type of fungi that form a mutualistic association with a plant root. The fungus extracts carbohydrate from the plant root while increasing the uptake of phosphorus, other mineral nutrients, and water for the plant.

Native species Taxa considered to have their origins in a given region or that have arrived there without recent (direct or indirect) transport by humans. Debate exists over how precisely to define this term.

Natural regeneration (also Spontaneous regeneration, Passive restoration) An approach to restoration that relies on spontaneous increases in biota without direct reintroduction after the removal of degrading factors alone, as distinct from an *assisted natural regeneration* or *active restoration/reconstruction* approach.

Natural resource managers Individuals responsible for managing land or water resources so that they are conserved for future generations.

Nitrogen-fixing plant species The process by which mutualistic bacteria convert atmospheric nitrogen into a form usable by plants. Some plants form mutualisms with nitrogen-fixing bacteria, plants such as many members of the pea family (Fabaceae), and a small number of other plant families.

Nonnative species (also Alien species, Exotic species) Taxa that do not have their origins in a given region and that have arrived there by recent (direct or indirect) transport by humans.

Nonpoint source pollutants   Pollutants that enter the water, air, or soil from diffuse sources, such as the case of excess fertilizer runoff from agricultural fields entering a river.

Novel ecosystems   Nonhistorical or novel species assemblages (i.e., species combinations and relative abundances that have not been observed in recent human history) due to anthropogenic environmental changes, land conversion, species invasions, extinctions, or a combination of these factors.

Nurse plant   A plant that *facilitates* the establishment of other plants, by various mechanisms such as attracting seed dispersers, increasing nutrient availability, or ameliorating stressful microclimatic conditions.

Objectives (also Performance criteria)   Specific outcomes needed to achieve the *goals* relative to any distinct spatial zones within the site. Objectives are stated in terms of measurable and quantifiable *parameters* to be able to evaluate whether they are being reached within a specified time.

Opportunity costs   The lost potential economic gain that occurs when a restoration site or certain resources provided by the site are not used for income-generating activities.

Organic matter   The pool of carbon-based compounds in the soil, usually formed through decomposition processes.

Parameter   A variable that is monitored to determine whether project *objectives* have been met.

Passive restoration   *See* Natural regeneration.

Payments for ecosystem services   Payments to farmers or landowners for managing their land to provide one or more *ecosystem services*.

Point bar   The deposited sediments that form along the edge of a winding arm of a river or stream, often forming an open, beach-like area.

Point source pollutants   Pollutants that enter the water, air, or soil from a single point. Examples include smokestacks and factory wastewater drains.

Practitioner   An individual who applies practical skills and knowledge to plan, implement, and monitor ecological restoration tasks.

Propagule   Any material that functions in propagating an organism (e.g., egg, seed, or clonal fragment). Propagules are produced by plants, fungi, bacteria, and animals.

Productivity   The rate of generation of biomass in an ecosystem, contributed to by the growth and reproduction of plants and animals.

Reclamation   Making severely degraded land (e.g., former mine sites or wastelands) fit for cultivation or a state suitable for some human use. There is not necessarily any native reference model defined or used; instead, emphasis is given to returning the site to an anthropocentrically useful condition or trajectory that provides desired ecosystem services.

Recovery   The process by which an ecosystem regains its *composition, structure*, and *processes* relative to the levels identified for the *reference model*. Usually follows a sequence of development called a *trajectory*. Restoration actions aim to assist the recovery process.

Recruitment   Production of a subsequent generation of organisms. Measured not by numbers of new organisms alone (e.g., not every hatchling or seedling) but by the number that survive as independent individuals in the population.

Reference model (also Reference ecosystem)   A native ecosystem that serves as a model for ecological restoration, which is informed by various sources of information often including one or multiple *reference sites*. A reference model usually represents a nondegraded version of the ecosystem complete with its *biota*, *abiotic* elements, *processes*, and *successional* states that would have existed on the restoration site had degradation not occurred—but adjusted to accommodate changed or predicted environmental conditions.

Reference site   An extant intact ecosystem representing attributes and a successional phase similar to project *objectives* against which progress at a restoration site can be compared over time through formal monitoring. Ideally, this monitoring involves more than one reference site.

Reforestation   Planting trees on land that was previously forested. The species used may or may not be native. This intervention may be undertaken as part of long-term forest restoration activity or for a specific use such as tree farming, *carbon storage*, or agroforestry.

Refugia   Areas in which a population of organisms can survive through a period of unfavorable conditions. Often used in the context of climate change.

Regeneration   *See* Natural regeneration, Assisted regeneration.

Reconstruction   *See* Active restoration.

Rehabilitation   Actions that aim to reinstate a level of ecosystem functionality where the goal is not *ecological restoration*, but rather the focus is on the provision of *ecosystem services*.

Reintroduction   Returning *biota* to an area where it previously occurred.

Restoration   *See* Ecological restoration.

Restoration ecology The branch of science that provides concepts, models, methodologies, and tools for the practice of *ecological restoration*. It also benefits from direct observation of and participation in restoration practice.

Revegetation Establishment, by any means, of plants on sites (including terrestrial, freshwater, and marine areas) that may or may not involve local or *native species*.

Rewilding Restoring an area of land to its uncultivated or "wild" state. Used especially with reference to the reintroduction of species of wild animal that have been driven out or exterminated so as to restore the processes that they affected (e.g., seed dispersal, grazing).

Riffles The shallow and rocky portion of a stream or river over which water typically passes quickly. These areas are biologically important for water aeration and faunal feeding.

Riparian Refers to the zone of contact or interface between land and a flowing surface water body, usually a river.

Ripping (of soil) Mechanically breaking up the soil surface with hook-shaped tines to reduce soil compaction and increase water infiltration.

Scarification (of seeds) The process of breaking the seed coat by chemical (e.g., through acid treatment) or mechanical (e.g., with abrasion or nicking) means to enhance seed germination of certain plant species.

Seed dormancy The physiological process by which seeds will not germinate, even when exposed to favorable germination conditions, due to chemical, physical, or other mechanisms that prevent germination until particular cues are applied.

Seed viability Viable seeds are capable of germinating under suitable conditions.

Sinuosity (of a river) The measure of how winding or curved a river or stream is, as measured by the total distance traveled by the water divided by the straight-line distance between two points along the flow.

Shifting baselines The phenomenon whereby each successive generation assumes that the diminished biological state is the norm, rather than recognizing that this state has itself been altered by prior human activities.

Soft release Providing time and support for animals to become conditioned to a new site (e.g., holding them in pens on site for a period, providing food for a time after they have been released).

Soil seed bank The stock of viable seeds, spores, and other plant *propagules* in the soil. The longevity of a seed bank varies from one or a few years to several decades or more depending on the species involved.

Soil texture   A measure of the relative percentages of sand, silt, and clay particles in a given soil. Soil texture affects water and air movement through the soil and processes such as nutrient *cycling*.

Soil compaction   A form of soil *degradation* in which soil particles are pressed into a smaller volume, resulting in less pore space for air or water, decreased water infiltration, and increased surface runoff.

Spatial heterogeneity   The patchy or uneven distribution of resources and species within a restoration site, watershed, or region. Small-scale spatial heterogeneity is typical of many natural ecosystems.

Species of concern   Species with declining populations that require conservation and restoration actions to persist. Here it is used to refer to both species that do and do not have legal protection under *endangered species* laws of some countries.

Stakeholders   All people and organizations who are involved in or affected by an action or policy and may be directly or indirectly included in the decision-making process; in restoration planning, stakeholders typically include government representatives, businesses, nongovernmental organizations, scientists, landowners, and local users of natural resources.

Stockpiling soil   The removal and short-term storage of soil to be replaced or reused elsewhere on site or in another project with the goal of maintaining the seed bank.

Stratification (of seeds)   The process of breaking dormancy by exposing seeds to cold temperatures for several weeks or months prior to sowing.

Subsidy (financial)   A financial benefit given to an individual or a business, usually in the form of a cash payment or tax reduction, to promote a certain sort of economic activity (e.g., agriculture).

Succession (ecological)   Patterns of change and replacement occurring within *ecosystems* over time in response to disturbance or lack of *disturbance*.

Surveillance monitoring   Monitoring with the explicit goal of watching for change and detecting unanticipated issues, rather than to evaluate whether a specific objective is met.

Swale   A low area of land that is used as an infiltration basin to increase rainwater infiltration and filter pollutants.

Target   *See* Goals and Objectives.

Topography   The arrangement and shape of the physical features on the surface of the earth, such as hills, valleys, and rivers.

Traditional cultural ecosystems Ecosystems that have developed under the joint influence of natural processes and human-imposed organization to provide *community composition, ecosystem structure*, and *ecosystem processes* more useful to human exploitation. Those considered high-quality examples of native ecosystems are able to function as *reference models* for *ecological restoration*, whereas others converted primarily to *nonnative species* or are otherwise modified from the traditional cultural ecosystem do not function as reference models.

Traditional ecological knowledge Knowledge and practices passed from generation to generation informed by strong cultural memories, sensitivity to change, and values that include reciprocity.

Trajectory (ecological) The course or pathway of an ecosystem over time. It may entail *degradation*, stasis, *natural regeneration*, adaptation to changing environmental conditions, or response to ecological restoration, ideally leading to recovery of *community composition* and *ecosystem processes*.

Translocation The intentional transport of organisms by humans to a different part of a given landscape or aquatic environment or to more distant areas. The purpose is generally to conserve an *endangered species*, subspecies, or population.

Vegetative propagation Growing new plants using vegetative parts, such as branches, stems, and roots, from other plants.

Waterbars Human-made structures built along trails or roads to slow and redirect water flow and reduce erosion.

Watershed The area of land that drains to a common waterway, such as a stream, lake, estuary, wetland, or aquifer, and from there into the nearest ocean.

Weathering The breakdown of rocks at Earth's surface by various physical, chemical, and biological processes.

References

Abdullah, M. M., R. A. Feagin, L. Musawi, S. Whisenant, and S. Popescu. 2016. "The use of remote sensing to develop a site history for restoration planning in an arid landscape." *Restoration Ecology* 24:91–99.

Alexander, C. A., F. Poulsen, D. C. Robinson, B. O. Ma, and R. A. Luster. 2018. "Improving multi-objective ecological flow management with flexible priorities and turn-taking: A case study from the Sacramento River and Sacramento–San Joaquin Delta." *San Francisco Estuary and Watershed Science* 16 (1): article 2. https://doi.org/10.15447/sfews.2018v16iss1/art2.

Alexander, J. M., and C. M. D'Antonio. 2003. "Control methods for the removal of French and Scotch Broom tested in coastal California." *Ecological Restoration* 21: 191–98.

Allen, E. B., M. E. Allen, L. Egerton-Warburton, L. Corkidi, and A. Gomez-Pompa. 2003. "Impacts of early- and late-seral mycorrhizae during restoration in seasonal tropical forest, Mexico." *Ecological Applications* 13:1701–17.

American Rivers. n.d. "Restoring damaged rivers." Accessed April 3, 2019. https://www.americanrivers.org/threats-solutions/restoring-damaged-rivers/.

Aronson, J., S. Dhillion, and E. Le Floch. 1995. "On the need to select an ecosystem of reference, however imperfect: A reply to Pickett and Parker." *Restoration Ecology* 3:1–3.

Aronson, J., S. J. Milton, and J. N. Blignaut. 2007. *Restoring Natural Capital*. Washington, DC: Island Press.

Axelrod, D. I. 1985. "Rise of the grassland biome, central North America." *Botanical Review* 51:163–201.

Backstrom, A. C., G. E. Garrard, R. J. Hobbs, and S. A. Bekessy. 2018. "Grappling with the social dimensions of novel ecosystems." *Frontiers in Ecology and the Environment* 16:109–17.

Baer, S. G. 2016. "Nutrient dynamics as determinants and outcomes of restoration." In *Foundations of Restoration Ecology*, 2nd ed., edited by M. A. Palmer, J. B. Zedler, and D. A. Falk, 333–64. Washington, DC: Island Press.

Bainbridge, D. A. 2012. *A Guide for Desert and Dryland Restoration: New Hope for Arid Lands*. Washington, DC: Island Press.

Bakker, J. D., E. G. Delvin, P. W. Dunwiddie, and J. Firn. 2018. "Staged-scale restoration:

Refining adaptive management to improve restoration effectiveness." *Journal of Applied Ecology* 55:1126–32.

Banes, G. L., B. M. F. Galdikas, and L. Vigilant. 2016. "Reintroduction of confiscated and displaced mammals risks outbreeding and introgression in natural populations, as evidenced by orangutans of divergent subspecies." *Scientific Reports* 6:article 22026. https://doi.org/10.1038/srep22026.

Barak, R. S., A. L. Hipp, J. Cavender-Bares, W. D. Pearse, S. C. Hotchkiss, E. A. Lynch, J. C. Callaway, et al. 2016. "Taking the long view: Integrating recorded, archeological, paleoecological, and evolutionary data into ecological restoration." *International Journal of Plant Sciences* 177:90–102.

Bayraktarov, E., M. I. Saunders, S. Abdullah, M. Mills, J. Beher, H. P. Possingham, P. J. Mumby, et al. 2016. "The cost and feasibility of marine coastal restoration." *Ecological Applications* 26:1055–74.

Beck, J., R. Carle, D. Calleri, and M. Hester. 2015. "Año Nuevo State Park seabird conservation and habitat restoration: Report 2015." Accessed April 29, 2019. http://oikonos.org/wp-content/uploads/2016/08/2015-Ano-Nuevo-Island-Seabird-Con sevation-Report_reduced-size.pdf.

BenDor, T. K., A. Livengood, T. W. Lester, A. Davis, and L. Yonavjak. 2015. "Defining and evaluating the ecological restoration economy." *Restoration Ecology* 23:209–19.

BenDor, T. K., J. Sholtes, and M. W. Doyle. 2009. "Landscape characteristics of a stream and wetland mitigation banking program." *Ecological Applications* 19:2078–92.

Berger, J. J. 1985. *Restoring the Earth*. New York: Knopf.

Bernhardt, E. S., M. A. Palmer, J. D. Allan, G. Alexander, K. Barnas, S. Brooks, J. Carr, et al. 2005. "Synthesizing US river restoration efforts." *Science* 308:636–37.

Beschta, R. L., and W. J. Ripple. 2016. "Riparian vegetation recovery in Yellowstone: The first two decades after wolf reintroduction." *Biological Conservation* 198:93–103.

Bonner, M. T. L., J. Herbohn, N. Gregorio, A. Pasa, M. S. Avela, C. Solano, M. O. M. Moreno, et al. 2019. "Soil organic carbon recovery in tropical tree plantations may depend on restoration of soil microbial composition and function." *Geoderma* 353:70–80.

Boonstra, F. 2010. "Leading by example: A comparison of New Zealand's and the United States' invasive species policies." *Connecticut Law Review* 43:1185–220.

Bower, A. D., B. St. Clair, and V. Erickson. 2014. "Generalized provisional seed zones for native plants." *Ecological Applications* 24:913–19.

Boyer, K. E., and J. B. Zedler. 1998. "Effects of nitrogen additions on the vertical structure of a constructed cordgrass marsh." *Ecological Applications* 8:692–705.

Bradshaw, A. D. 1984. "Land restoration: Now and in the future." *Proceedings of the Royal Society of London B* 223:1–23.

Bradshaw, A. D. 1987. "Restoration: An acid test for ecology." In *Restoration Ecology*, edited by W. R. Jordan III, M. Gilpin, and J. D. Aber, 23–29. Cambridge: Cambridge University Press.

Bradshaw, A. D., and M. J. Chadwick. 1980. *The Restoration of Land*. Berkeley: University of California Press.

Brancalion, P. H. S., C. Bello, R. L. Chazdon, M. Galetti, P. Jordano, R. A. F. Lima, A. Medina, M. A. Pizo, and J. L. Reid. 2018. "Maximizing biodiversity conservation and carbon stocking in restored tropical forests." *Conservation Letters* 11: article e12454. https://onlinelibrary.wiley.com/doi/abs/10.1111/conl.12454.

Brancalion, P. H. S., L. C. Garcia, R. Loyola, R. R. Rodrigues, V. D. Pillar, and T. M. Lewinsohn. 2016. "A critical analysis of the Native Vegetation Protection Law of Brazil (2012): Updates and ongoing initiatives." *Natureza and Conservação* 14, Supplement 1:1–15.

Brancalion, P. H. S., D. Lamb, E. Ceccon, D. Boucher, J. Herbohn, B. Strassburg, and D. P. Edwards. 2017. "Using markets to leverage investment in forest and landscape restoration in the tropics." *Forest Policy and Economics* 85:103–13.

Brancalion, P. H. S., R. A. G. Viani, J. Aronson, R. R. Rodrigues, and A. G. Nave. 2012. "Improving planting stocks for the Brazilian Atlantic forest restoration through community-based seed harvesting strategies." *Restoration Ecology* 20:704–11.

Brancalion, P. H. S., R. A. G. Viani, B. B. N. Strassburg, and R. R. Rodrigues. 2012. "Finding the money for tropical forest restoration." *Unasylva* 63:41–49.

Breed, M. F., M. G. Stead, K. M. Ottewell, M. G. Gardner, and A. J. Lowe. 2013. "Which provenance and where? Seed sourcing strategies for revegetation in a changing environment." *Conservation Genetics* 14:1–10.

Bremer, L. L., D. A. Auerbach, J. H. Goldstein, A. L. Vogl, D. Shemie, T. Kroeger, J. L. Nelson, et al. 2016. "One size does not fit all: Natural infrastructure investments within the Latin American Water Funds Partnership." *Ecosystem Services* 17:217–36.

Brierley, G. J., and K. Fryirs. 2008. *River Futures*. Washington, DC: Island Press.

Briske, D. D., S. D. Fuhlendorf, and E. E. Smeins. 2005. "State-and-transition models, thresholds, and rangeland health: A syntheis of ecological concepts and perspectives." *Rangeland Ecology and Management* 58:1–11.

Britton, J. R., R. E. Gozlan, and G. H. Copp. 2011. "Managing non-native fish in the environment." *Fish and Fisheries* 12:256–74.

Brotons, L., N. Aquilué, M. de Cáceres, M.-J. Fortin, and A. Fall. 2013. "How fire history, fire suppression practices and climate change affect wildfire regimes in Mediterranean landscapes." *PLOS ONE* 8: article e62392. https://doi.org/10.1371/journal.pone.0062392.

Brown, P. H., and C. L. Lant. 1999. "The effect of wetland mitigation banking on the achievement of no-net-loss." *Environmental Management* 23:333–45.

Bugosh, N., and E. Epp. 2019. "Evaluating sediment production from native and fluvial geomorphic-reclamation watersheds at La Plata Mine." *CATENA* 174:383–98.

Cadenasso, M. L., S. T. A. Pickett, K. C. Weathers, and C. G. Jones. 2003. "A framework for a theory of ecological boundaries." *BioScience* 53:750–58.

Calle, Z., E. Murgueitio, J. Chará, C. H. Molina, A. F. Zuluaga, and A. Calle. 2013. "A strategy for scaling-up intensive silvopastoral systems in Colombia." *Journal of Sustainable Forestry* 32:677–93.

Carbyn, L. N., H. J. Armbruster, and C. Mamo. 1994. "The swift fox reintroduction program in Canada from 1983 to 1992." In *Restoration of Endangered Species: Conceptual Issues, Planning, and Implementation*, edited by M. L. Bowles and C. J. Whelan, 247–71. Cambridge: Cambridge University Press.

Carey, M. P., B. L. Sanderson, K. A. Barnas, and J. D. Olden. 2012. "Native invaders—Challenges for science, management, policy, and society." *Frontiers in Ecology and the Environment* 10:373–81.

Center for Invasive Species and Ecosystem Health. n.d. "Early detection and distribution mapping system." Accessed January 24, 2019. https://www.eddmaps.org/.

César, R. G., K. D. Holl, V. J. Girão, F. N. A. Mello, E. Vidal, M. C. Alves, and P. H. S.

Brancalion. 2016. "Evaluating climber cutting as a strategy to restore degraded tropical forests." *Biological Conservation* 201:309–13.

Chapman, M. G. 1999. "Improving sampling designs for measuring restoration in aquatic habitats." *Journal of Aquatic Ecosystem Stress and Recovery* 6:235–51.

Chaves, R. B., G. Durigan, P. H. S. Brancalion, and J. Aronson. 2015. "On the need of legal frameworks for assessing restoration projects success: New perspectives from São Paulo state (Brazil)." *Restoration Ecology* 23:754–59.

Chazdon, R. L., and M. R. Guariguata. 2016. "Natural regeneration as a tool for large-scale forest restoration in the tropics: Prospects and challenges." *Biotropica* 48:716–30.

Chechina, M., and A. Hamann. 2015. "Choosing species for reforestation in diverse forest communities: Social preference versus ecological suitability." *Ecosphere* 6: article 240. http://dx.doi.org/10.1890–0S15-00131.1.

Chiquoine, L. P., S. R. Abella, and M. A. Bowker. 2016. "Rapidly restoring biological soil crusts and ecosystem functions in a severely disturbed desert ecosystem." *Ecological Applications* 26:1260–72.

Clements, F. E. 1916. *Plant Successon and Indicators*. New York: Henry Wilson Co.

Clewell, A. F., and J. Aronson. 2006. "Motivations for the restoration of ecosystems." *Conservation Biology* 20:420–28.

Clewell, A. F., and J. Aronson. 2013. *Ecological Restoration: Principles, Values, and Structure of an Emerging Profession*. Washington, DC: Island Press.

Cliquet, A. 2017. "International law and policy on restoration." In *Routledge Handbook of Ecological and Environmental Restoration*, edited by S. K. Allison and S. D. Murphy, 381–400. London: Routledge.

Collier, N., B. J. Austin, C. J. A. Bradshaw, and C. R. McMahon. 2011. "Turning pests into profits: Introduced buffalo provide multiple benefits to indigenous people of Northern Australia." *Human Ecology* 39:155–64.

Collinge, S. K., C. Ray, and F. Gerhardt. 2011. "Long-term dynamics of biotic and abiotic resistance to exotic species invasion in restored vernal pool plant communities." *Ecological Applications* 21:2105–18.

Conner, R., D. C. Rudolph, and J. R. Walters. 2001. *The Red-Cockaded Woodpecker: Surviving in a Fire-Maintained Ecosystem*. Austin: University of Texas Press.

Cooke, G. D., E. B. Welch, S. Peterson, and S. A. Nichols. 2016. *Restoration and Management of Lakes and Reservoirs*. Boca Raton, FL: CRC Press.

Corbin, J. D., and K. D. Holl. 2012. "Applied nucleation as a forest restoration strategy." *Forest Ecology and Management* 265:37–46.

Corlett, R. T. 2016. "Restoration, reintroduction, and rewilding in a changing world." *Trends in Ecology and Evolution* 31:453–62.

Costanza, R., and L. Cornwell. 1992. "The 4P approach to dealing with scientific uncertainty." *Environment: Science and Policy for Sustainable Development* 34:12–42.

Craft, C. 2016. *Creating and Restoring Wetlands: From Theory to Practice*. Amsterdam: Elsevier.

Crivelli, A. J. 1995. "Are fish introductions a threat to endemic freshwater fishes in the northern Mediterranean region?" *Biological Conservation* 72:311–19.

Crowley, S. L., S. Hinchliffe, and R. A. McDonald. 2017. "Invasive species management will benefit from social impact assessment." *Journal of Applied Ecology* 54:351–57.

Crutzen, P. J. 2002. "Geology of mankind." *Nature* 415:23.

Cuthbert, R. J., A. M. Taggart, V. Prakash, S. S. Chakraborty, P. Deori, T. Galligan, M. Kulkarni, et al. 2014. "Avian scavengers and the threat from veterinary pharmaceuticals." *Philosophical Transactions of the Royal Society B: Biological Sciences* 369: article 20130574.

Danielsen, F., M. Skutsch, N. D. Burgess, P. M. Jensen, H. Andrianandrasana, B. Karky, R. Lewis, et al. 2011. "At the heart of REDD+: A role for local people in monitoring forests?" *Conservation Letters* 4:158–67.

D'Antonio, C. M., E. August-Schmidt, and B. Fernandez-Going. 2016. "Invasive species and restoration challenges." In *Foundations of Restoration Ecology*, 2nd ed., edited by M. A. Palmer, J. B. Zedler, and D. A. Falk, 216–44. Washington, DC: Island Press.

D'Antonio, C. M., and P. M. Vitousek. 1992. "Biological invasions by exotic grasses, the grass/fire cycle, and global change." *Annual Review of Ecology and Systematics* 23:63–87.

Davidson, N. C. 2014. "How much wetland has the world lost? Long-term and recent trends in global wetland area." *Marine and Freshwater Research* 65:934–41.

Davis, M. A., M. K. Chew, R. J. Hobbs, A. E. Lugo, J. J. Ewel, G. J. Vermeij, J. H. Brown, et al. 2011. "Don't judge species on their origins." *Nature* 474:153–54.

De Groot, R. S., J. Blignaut, S. Van Der Ploeg, J. Aronson, T. Elmqvist, and J. Farley. 2013. "Benefits of investing in ecosystem restoration." *Conservation Biology* 27:1286–93.

Derak, M., J. Cortina, L. Taiqui, and A. Aledo. 2018. "A proposed framework for participatory forest restoration in semiarid areas of North Africa." *Restoration Ecology* 26:S18–S25.

Destro, G. F. G., P. De Marco, and L. C. Terribile. 2018. "Threats for bird population restoration: A systematic review." *Perspectives in Ecology and Conservation* 16:68–73.

Ding, H., S. Faruqi, A. Wu, J. C. Altamirano, A. Anchondo Ortega, M. Verdone, R. Zamora Cristales, et al. 2017. *Roots of Prosperity*. Washington, DC: World Resources Institute.

Docker, B., and I. Robinson. 2014. "Environmental water management in Australia: Experience from the Murray-Darling Basin." *International Journal of Water Resources Development* 30:164–77.

Doherty, J. M., J. F. Miller, S. G. Prellwitz, A. M. Thompson, S. P. Loheide, and J. B. Zedler. 2014. "Hydrologic regimes revealed bundles and tradeoffs among six wetland services." *Ecosystems* 17:1026–39.

Doherty, J. M., and J. B. Zedler. 2015. "Increasing substrate heterogeneity as a bet-hedging strategy for restoring wetland vegetation." *Restoration Ecology* 23:15–25.

Dolan, R. W., K. A. Harris, and M. Adler. 2015. "Community involvement to address a long-standing invasive species problem: Aspects of civic ecology in practice." *Ecological Restoration* 33:316–25.

Dörnhöfer, K., and N. Oppelt. 2016. "Remote sensing for lake research and monitoring—Recent advances." *Ecological Indicators* 64:105–22.

Drexler, J. Z., I. Woo, C. C. Fuller, and G. Nakai. 2019. "Carbon accumulation and vertical accretion in a restored versus historic salt marsh in southern Puget Sound, Washington, United States." *Restoration Ecology* 27. https://doi.org/10.1111/rec .12941.

Dudley, T. L., and D. W. Bean. 2012. "Tamarisk biocontrol, endangered species risk and resolution of conflict through riparian restoration." *BioControl* 57:331–47.

Durigan, G., N. Guerin, and J. da Costa. 2013. "Ecological restoration of Xingu Basin headwaters: Motivations, engagement, challenges and perspectives." *Philosophical Transactions of the Royal Society B-Biological Sciences* 368: article 20120165. https://doi.org/10.1098/rstb.2012.0165.

East, A. 2017. "They released 14 wolves in a park. But no one was prepared for this." Accessed March 27, 2018. https://weloveanimals.me/released-14-wolves-park-no-one-prepared-unbelievable-nature11/.

Egan, D., E. E. Hjerpe, and J. Abrams (eds.). 2011. *Human Dimensions of Ecological Restoration.* Washington, DC: Island Press.

Egan, D., and E. A. Howell. 2001. *The Historical Ecology Handbook: A Restorationist's Guide to Reference Ecosystems.* Washington, DC: Island Press.

ELD Initiative. 2015. "The value of land: Prosperous lands and positive rewards through sustainable land management." Accessed November 13, 2018. http://www.eld-initiative.org/fileadmin/pdf-0LD-main-report_05_web_72dpi.pdf.

Elliott, S. 2016. "The potential for automating assisted natural regeneration of tropical forest ecosystems." *Biotropica* 48:825–33.

Elzinga, C. L., D. W. Salzer, and J. W. Willoughby. 1998. *Measuring and Monitoring Plant Populations.* Denver, CO: Bureau of Land Management.

Eschen, R., K. Britton, E. Brockerhoff, T. Burgess, V. Dalley, R. S. Epanchin-Niell, K. Gupta, et al. 2015. "International variation in phytosanitary legislation and regulations governing importation of plants for planting." *Environmental Science and Policy* 51:228–37.

Evans, K., M. R. Guariguata, and P. H. S. Brancalion. 2018. "Participatory monitoring to connect local and global priorities for forest restoration." *Conservation Biology* 32:525–34.

Ewers, R. M., and R. K. Didham. 2006. "Confounding factors in the detection of species responses to habitat fragmentation." *Biological Reviews* 81:117–42.

Falk, D. A. 2017. "Restoration ecology, resilience, and the axes of change." *Annals of the Missouri Botanical Garden* 102:201–16.

FAO. n.d. "Agricultural land (% of land area)." Accessed November 13, 2018. https://data.worldbank.org/indicator/AG.LND.AGRI.ZS.

Ferrario, F., M. W. Beck, C. D. Storlazzi, F. Micheli, C. C. Shepard, and L. Airoldi. 2014. "The effectiveness of coral reefs for coastal hazard risk reduction and adaptation." *Nature Communications* 5: article 3794. https://www.nature.com/articles/ncomms4794.

Ferren, W. R., Jr., D. M. Hubbard, S. Wiseman, A. K. Parikh, and N. Gale. 1998. "Review of ten years of vernal pool restoration and creation in Santa Barbara, California." In *Ecology, Conservation, and Management of Vernal Pool Ecosystems: Proceedings from a 1996 Conference,* edited by C. W. Witham, E. T. Bauder, D. Belk, W. R. Ferren, Jr., and R. Ornduff, 206–16. Sacramento, CA: California Native Plant Society.

Feyera, S., E. Beck, and U. Lüttge. 2002. "Exotic trees as nurse-trees for the regeneration of natural tropical forests." *Trees* 16:245–49.

Filoso, S., M. O. Bezerra, K. C. B. Weiss, and M. A. Palmer. 2017. "Impacts of forest restoration on water yield: A systematic review." *PLOS ONE* 12: article e0183210. https://doi.org/10.1371/journal.pone.0183210.

Firestone, J., and J. J. Corbett. 2005. "Coastal and port environments: International legal and policy responses to reduce ballast water introductions of potentially invasive species." *Ocean Development and International Law* 36:291–316.

Fischer, J., and D. B. Lindenmayer. 2000. "An assessment of the published results of animal relocations." *Biological Conservation* 96:1–11.

Forman, R. T. T., and M. Godron. 1981. "Patches and structural components for a landscape ecology." *BioScience* 31:733–40.

Fritts, T. H., and G. H. Rodda. 1998. "The role of introduced species in the degradation of island ecosystems: A case history of Guam." *Annual Review of Ecology and Systematics* 29:113–40.

Funk, J. L., E. E. Cleland, K. N. Suding, and E. S. Zavaleta. 2008. "Restoration through reassembly: Plant traits and invasion resistance." *Trends in Ecology and Evolution* 23:695–703.

Galatowitsch, S. M., and J. B. Zedler. 2014. "Wetland restoration." In *Ecology of Freshwater and Estuarine Wetlands*, edited by D. P. Batzer and R. R. Sharitz, 225–60. Berkeley: University of California Press.

Gann, G. D., T. McDonald, B. Walder, J. Aronson, C. R. Nelson, J. Jonson, C. Eisenberg, et al. 2019. *International Principles and Standards for the Practice of Ecological Restoration.* Washington, DC: Society for Ecological Restoration.

Garcia, C. 2017. "Supporting conservation by playing a game? Seriously." Accessed November 26, 2018. https://news.mongabay.com/2017/10/supporting-conservation-by-playing-a-game-seriously-commentary/.

Geist, H., W. McConnell, E. F. Lambin, E. Moran, D. Alves, and T. Rudel. 2006. "Causes and trajectories of land-use/cover change." In *Land-use and Land-cover Change*, edited by H. J. Geist and E. F. Lambin, 41–70. Berlin: Springer.

Gelfenbaum, G., A. W. Stevens, I. Miller, J. A. Warrick, A. S. Ogston, and E. Eidam. 2015. "Large-scale dam removal on the Elwha River, Washington, USA: Coastal geomorphic change." *Geomorphology* 246:649–68.

Gerard, D. 2000. "The law and economics of reclamation bonds." *Resources Policy* 26:189–97.

Glenn, E. P., P. L. Nagler, P. B. Shafroth, and C. J. Jarchow. 2017. "Effectiveness of environmental flows for riparian restoration in arid regions: A tale of four rivers." *Ecological Engineering* 106:695–703.

Gomez-Aparicio, L. 2009. "The role of plant interactions in the restoration of degraded ecosystems: A meta-analysis across life-forms and ecosystems." *Journal of Ecology* 97:1202–14.

Goosem, S., and N. I. G. Tucker. 2013. *Repairing the Rainforest.* Cairns: Wet Tropics Management Authority and Biotropica Australia.

Greco, S. E. 1999. "Monitoring riparian landscape change and modeling habitat dynamics of the yellow-billed cuckoo on the Sacramento River, California." PhD diss., University of California, Davis.

Greenbelt Movement. n.d. "Our history." Accessed November 27, 2018. https://www.greenbeltmovement.org/who-we-are/our-history.

Greene, H. C., and J. T. Curtis. 1950. "Germination studies of Wisconsin prairie plants." *American Midland Naturalist* 39:186–94.

Greene, H. C., and J. T. Curtis. 1953. "The re-establishment of prairie in the University of Wisconsin Arboretum." *Wild Flower* 29:77–88.

Griffin, C. P. 1998. *Factors Affecting Captive Prairie Chicken Production.* PhD diss., Texas A&M. University.

Grootjans, A. P., H. W. T. Geelen, A. J. M. Jansen, and E. J. Lammerts. 2002. "Restoration of coastal dune slacks in the Netherlands." In *Ecological Restoration of Aquatic and Semi-Aquatic Ecosystems in the Netherlands (NW Europe)*, edited by P. H. Nienhuis and R. D. Gulati, 181–203. Dordrecht: Springer Netherlands.

Guerrero, A. M., L. Shoo, G. Iacona, R. J. Standish, C. P. Catterall, L. Rumpff, K. de Bie, et al. 2017. "Using structured decision-making to set restoration objectives when multiple values and preferences exist." *Restoration Ecology* 25:858–65.

Gulati, R. D., L. M. D. Pires, and E. van Donk. 2012. "Restoration of freshwater lakes." In *Restoration Ecology*, edited by J. Van Andel and J. Aronson, 233–47. Malden, MA: Blackwell.

Gunn, J. 1995. *Restoration and Recovery of an Industrial Region—Progress in Restoring the Smelter-Damaged Landscape near Sudbury, Canada*. New York: Springer-Verlag.

Gutierrez, V., and M. Keijzer. 2015. "Funding forest landscape restoration using a business-centred approach: An NGO's perspective." *Unasylva* 66:99.

Hale, R., R. Mac Nally, D. T. Blumstein, and S. E. Swearer. 2019. "Evaluating where and how habitat restoration is undertaken for animals." *Restoration Ecology* 27: 775–81.

Hallett, L. M., R. J. Standish, K. B. Hulvey, M. R. Gardener, K. N. Suding, B. M. Starzomski, S. D. Murphy, et al. 2013. "Towards a conceptual framework for novel ecosystems." In *Novel Ecosystems*, edited by R. J. Hobbs, E. S. Higgs and C. M. Hall, 17–28. Hoboken, NJ: Wiley.

Hammond, B. W. 1999. "*Saccharum spontaneum* (Gramineae) in Panama." *Journal of Sustainable Forestry* 8:23–38.

Hansen, A. T., C. L. Dolph, E. Foufoula-Georgiou, and J. C. Finlay. 2018. "Contribution of wetlands to nitrate removal at the watershed scale." *Nature Geoscience* 11:127–32.

Hansson, A., and P. Dargusch. 2017. "An estimate of the financial cost of peatland restoration in Indonesia." *Case Studies in the Environment*. http://cse.ucpress.edu/content/ecs/early/2017/12/17/cse.2017.000695.full.pdf.

Havens, K., P. Vitt, S. Still, A. T. Kramer, J. B. Fant, and K. Schatz. 2015. "Seed sourcing for restoration in an era of climate change." *Natural Areas Journal* 35:122–33.

Hawlena, D., D. Saltz, Z. Abramsky, and A. Bouskila. 2010. "Ecological trap for desert lizards caused by anthropogenic changes in habitat structure that favor predator activity." *Conservation Biology* 24:803–9.

Hedrick, P. W., and R. Fredrickson. 2010. "Genetic rescue guidelines with examples from Mexican wolves and Florida panthers." *Conservation Genetics* 11:615–26.

Herman, M. R., and A. P. Nejadhashemi. 2015. "A review of macroinvertebrate- and fish-based stream health indices." *Ecohydrology and Hydrobiology* 15:53–67.

Herrick, J. E., J. W. Van Zee, K. Havstad, L. M. Burkett, and W. G. Whitford. 2005. *Monitoring Manual for Grassland, Shrubland and Savanna Ecosystems*. Las Cruces, NM: USDA–ARTS Jornada Experimental Range.

Higgs, E., D. A. Falk, A. Guerrini, M. Hall, J. Harris, R. J. Hobbs, S. T. Jackson, et al. 2014. "The changing role of history in restoration ecology." *Frontiers in Ecology and the Environment* 12:499–506.

Hilderbrand, R. H., A. C. Watts, and A. M. Randle. 2005. "The myths of restoration ecology." *Ecology and Society* 10: article 19. http://www.ecologyandsociety.org/vol10/iss1/art19/.

Hobbs, R. J., L. M. Hallett, P. R. Ehrlich, and H. A. Mooney. 2011. "Intervention ecology: Applying ecological science in the twenty-first century." *BioScience* 61:442–50.

Hobbs, R. J., E. Higgs, and J. A. Harris. 2009. "Novel ecosystems: Implications for conservation and restoration." *Trends in Ecology and Evolution* 24:599–605.

Hobbs, R. J., and S. E. Humphries. 1995. "An integrated approach to the ecology and management of plant invasions." *Conservation Biology* 9:761–70.

Hobbs, R. J., and K. N. Suding. 2009. *New Models for Ecosystem Dynamics and Restoration.* Washington, DC: Island Press.

Hoddle, M. n.d. "Quagga and zebra mussels." Accessed January 2, 2018. http://cisr.ucr.edu/quagga_zebra_mussels.html.

Holl, K. D. 2002a. "Effect of shrubs on tree seedling establishment in abandoned tropical pasture." *Journal of Ecology* 90:179–87.

Holl, K. D. 2002b. "Long-term vegetation recovery on reclaimed coal surface mines in the eastern USA." *Journal of Applied Ecology* 39:960–70.

Holl, K. D. 2012. "Tropical forest restoration." In *Restoration Ecology*, edited by J. Van Andel and J. Aronson, 103–14. Malden, MA: Blackwell Publishing.

Holl, K. D., and T. M. Aide. 2011. "When and where to actively restore ecosystems?" *Forest Ecology and Management* 261:1558–63.

Holl, K. D., and J. Cairns Jr. 2002. "Monitoring and appraisal." In *Handbook of Ecological Restoration*, vol. 1, edited by M. R. Perrow and A. J. Davy, 411–32. Cambridge: Cambridge University Press.

Holl, K. D., E. E. Crone, and C. B. Schultz. 2003. "Landscape restoration: Moving from generalities to methodologies." *BioScience* 53:491–502.

Holl, K. D., E. A. Howard, T. M. Brown, R. G. Chan, T. S de Silva, E. T. Mann, J. A. Russell, et al. 2014. "Efficacy of exotic control strategies for restoring coastal prairie grasses." *Invasive Plant Science and Management* 7:590–98.

Holl, K. D., and R. D. Howarth. 2000. "Paying for restoration." *Restoration Ecology* 8:260–67.

Holloran, P., A. Mackenzie, S. Farrell, and D. Johnson. 2004. *The Weed Workers' Handbook.* Richmond, CA: California Invasive Plant Council.

Hua, F. Y., X. Y. Wang, X. L. Zheng, B. Fisher, L. Wang, J. G. Zhu, Y. Tang, et al. 2016. "Opportunities for biodiversity gains under the world's largest reforestation programme." *Nature Communications* 7: article 12717. https://www.nature.com/articles/ncomms12717.pdf.

Hüttermann, A., L. J. B. Orikiriza, and H. Agaba. 2009. "Application of superabsorbent polymers for improving the ecological chemistry of degraded or polluted lands." *Clean—Soil, Air, Water* 37:517–26.

Iftekhar, M. S., M. Polyakov, D. Ansell, F. Gibson, and G. M. Kay. 2017. "How economics can further the success of ecological restoration." *Conservation Biology* 31:261–68.

Intergovernmental Science-Policy Platform on Biodiversity and Ecosystem Services. 2018. *Summary for Policymakers of the Assessment Report on Land Degradation and Restoration of the Intergovernmental Science-Policy Platform on Biodiversity and Ecosystem Services.* Edited by R. J. Scholes, L. Montanarella, E. Brainich, N. Barger, B. ten Brink, M. Cantele, B. Erasmus, et al. Bonn: IPBES Secretariat.

International Rivers. 2014. "The state of the world's rivers." Accessed May 2, 2018. https://www.internationalrivers.org/resources/8391.

Inyo County Water Department. n.d. "Lower Owens River project." Accessed January 8, 2019. http://www.inyowater.org/projects/lorp/.

Island Conservation. 2017. "Impact report 2016/2017." Accessed March 26, 2018. https://www.islandconservation.org/report/2017/.

Jankowski, K. L., T. E. Törnqvist, and A. M. Fernandes. 2017. "Vulnerability of Louisiana's coastal wetlands to present-day rates of relative sea-level rise." *Nature Communications* 8: article 14792. http://dx.doi.org/10.1038/ncomms14792.

Jones, H. P., N. D. Holmes, S. H. M. Butchart, B. R. Tershy, P. J. Kappes, I. Corkery, A. Aguirre-Muñoz, et al. 2016. "Invasive mammal eradication on islands results in substantial conservation gains." *Proceedings of the National Academy of Sciences* 113:4033–38.

Jordan, W. R., III. 2003. *The Sunflower Forest.* Berkeley: University of California Press.

Kaiser-Bunbury, C. N., J. Mougal, A. E. Whittington, T. Valentin, R. Gabriel, J. M. Olesen, and N. Blüthgen. 2017. "Ecosystem restoration strengthens pollination network resilience and function." *Nature* 542:223–27.

Kark, S., A. Tulloch, A. Gordon, T. Mazor, N. Bunnefeld, and N. Levin. 2015. "Cross-boundary collaboration: Key to the conservation puzzle." *Current Opinion in Environmental Sustainability* 12:12–24.

Keeley, J. E. 2002. "Native American impacts on fire regimes of the California coastal ranges." *Journal of Biogeography* 29:303–20.

Keeley, J. E., and C. J. Fotheringham. 1998. "Smoke-induced seed germination in California chaparral." *Ecology* 79:2320–36.

Kerr, D. W., I. B. Hogle, B. S. Ort, and W. J. Thornton. 2016. "A review of 15 years of Spartina management in the San Francisco Estuary." *Biological Invasions* 18:2247–66.

Kimball, S., M. Lulow, Q. Sorenson, K. Balazs, Y.-C. Fang, S. J. Davis, M. O'Connell, and T. E. Huxman. 2015. "Cost-effective ecological restoration." *Restoration Ecology* 23:800–810.

Klimkowska, A., W. Kotowski, R. van Diggelen, A. P. Grootjans, P. Dzierza, and K. Brzezinska. 2010. "Vegetation re-development after fen meadow restoration by topsoil removal and hay transfer." *Restoration Ecology* 18:924–33.

Koch, J. M. 2007. "Restoring a jarrah forest understorey vegetation after bauxite mining in Western Australia." *Restoration Ecology* 15:S26–S39.

Koebel, J. W., and S. G. Bousquin. 2014. "The Kissimmee River restoration project and evaluation program, Florida, U.S.A." *Restoration Ecology* 22:345–52.

Kondolf, G. M. 1995. "Five elements for effective evaluation of stream restoration." *Restoration Ecology* 3:133–36.

Kronvang, B., L. Svendsen, A. Brookes, K. Fisher, B. Møller, O. Ottosen, M. Newson, et al. 1998. "Restoration of the rivers Brede, Cole and Skerne: A joint Danish and British EU–LIFE demonstration project, III—Channel morphology, hydrodynamics and transport of sediment and nutrients." *Aquatic Conservation: Marine and Freshwater Ecosystems* 8:209–22.

Ladouceur, E., B. Jiménez-Alfaro, M. Marin, M. De Vitis, H. Abbandonato, P. P. M. Iannetta, C. Bonomi, et al. 2018. "Native seed supply and the restoration species pool." *Conservation Letters* 11: article e12381. http://dx.doi.org/10.1111/conl.12381.

Lambin, E. F., and P. Meyfroidt. 2011. "Global land use change, economic globalization, and the looming land scarcity." *Proceedings of the National Academy of Sciences* 108:3465–72.

Lande, R. 1993. "Risks of population extinction from demographic and environmental stochasticity and random catastrophes." *American Naturalist* 142:911–27.

Larkin, D. J., G. L. Bruland, and J. B. Zedler. 2016. "Heterogeneity theory and ecological restoration." In *Foundations of Restoration Ecology*, 2nd ed., edited by M. A. Palmer, J. B. Zedler, and D. A. Falk, 271–300. Washington, DC: Island Press.

Latawiec, A. E., B. B. N. Strassburg, P. H. S. Brancalion, R. R. Rodrigues, and T. Gardner. 2015. "Creating space for large-scale restoration in tropical agricultural landscapes." *Frontiers in Ecology and the Environment* 13:211–18.

Leahy, J. G., and R. R. Colwell. 1990. "Microbial degradation of hydrocarbons in the environment." *Microbiological Reviews* 54:305–15.

Lees, A. C., and C. A. Peres. 2008. "Conservation value of remnant riparian forest corridors of varying quality for Amazonian birds and mammals." *Conservation Biology* 22:439–49.

Le Maitre, D. C., M. Gaertner, E. Marchante, E.-J. Ens, P. M. Holmes, A. Pauchard, P. J. O'Farrell, et al. 2011. "Impacts of invasive Australian acacias: Implications for management and restoration." *Diversity and Distributions* 17:1015–29.

Lesage, J. C., E. A. Howard, and K. D. Holl. 2018. "Homogenizing biodiversity in restoration: The 'perennialization' of California prairies." *Restoration Ecology* 26:1061–65.

Lindell, C. A. 2008. "The value of animal behavior in evaluations of restoration success." *Restoration Ecology* 16:197–203.

Lindenmayer, D. B., and G. E. Likens. 2018. *Effective Ecological Monitoring*. 2nd ed. London: Earthscan.

Locatelli, B., C. P. Catterall, P. Imbach, C. Kumar, R. Lasco, E. Marin-Spiotta, B. Mercer, et al. 2015. "Tropical reforestation and climate change: Beyond carbon." *Restoration Ecology* 23:337–43.

Lockwood, J. L., and S. L. Pimm. 1999. "When does restoration succeed?" In *Ecological Assembly Rules*, edited by E. Weiher and P. Keddy, 363–92. Cambridge: Cambridge University Press.

Lopes-Fernandes, M., C. Espírito-Santo, and A. Frazão-Moreira. 2018. "The return of the Iberian lynx to Portugal: Local voices." *Journal of Ethnobiology and Ethnomedicine* 14:3. https://doi.org/10.1186/s13002-017-0200-9.

Louda, S. M., and C. W. O'Brien. 2002. "Unexpected ecological effects of distributing the exotic weevil." *Conservation Biology* 16:717–27.

MacArthur, R. H., and E. O. Wilson. 1967. *The Theory of Island Biogeography*. Princeton, NJ: Princeton University Press.

Maheshwari, A., N. Midha, and A. Cherukupalli. 2014. "Participatory rural appraisal and compensation intervention: Challenges and protocols while managing large carnivore–human conflict." *Human Dimensions of Wildlife* 19:62–71.

Mamun, A.-A. 2010. "Understanding the value of local ecological knowledge and practices for habitat restoration in human-altered floodplain systems: A case from Bangladesh." *Environmental Management* 45:922–38.

Mansourian, S. 2017. "Governance and restoration." In *Routledge Handbook of Ecological and Environmental Restoration*, edited by S. K. Allison and S. D. Murphy, 401–13. London: Routledge.

Marin-Spiotta, E., and R. Ostertag. 2016. "Recovery of ecosystem processes: Carbon and energy flows in restored wetlands, grassland, and forests." In *Foundations of Restoration Ecology*, 2nd ed., edited by M. A. Palmer, J. B. Zedler, and D. A. Falk, 365–94. Washington, DC: Island Press.

Maron, M., C. D. Ives, H. Kujala, J. W. Bull, F. J. F. Maseyk, S. Bekessy, A. Gordon, et al. 2016. "Taming a wicked problem: Resolving controversies in biodiversity offsetting." *BioScience* 66:489–98.

Marsh, L. L., D. R. Porter, and D. A. Slaveson. 1996. *Mitigation Banking: Theory and Practice.* Washington, DC: Island Press.

Martin, D. M., and J. E. Lyons. 2018. "Monitoring the social benefits of ecological restoration." *Restoration Ecology* 26:1045–50.

Maschinski, J., and K. E. Haskins. 2012. *Plant Reintroduction in a Changing Climate: Promises and Perils.* Washington, DC: Island Press.

Matzek, V., C. Puleston, and J. Gunn. 2015. "Can carbon credits fund riparian forest restoration?" *Restoration Ecology* 23:7–14.

May, J., R. J. Hobbs, and L. E. Valentine. 2017. "Are offsets effective? An evaluation of recent environmental offsets in Western Australia." *Biological Conservation* 206: 249–57.

Mazaika, K. 2004. "'The Mono Lake case." In *Braving the Currents: Evaluating Environmental Conflict Resolution in the River Basins of the American West,* edited by T. P. d'Estrée and B. G. Colby, 71–105. New York: Springer US.

McAlpine, C., C. P. Catterall, R. Mac Nally, D. Lindenmayer, J. L. Reid, K. D. Holl, A. F. Bennett, et al. 2016. "Integrating plant- and animal-based perspectives for more effective restoration of biodiversity." *Frontiers in Ecology and the Environment* 14:37–45.

McCallum, K. P., A. J. Lowe, M. F. Breed, and D. C. Paton. 2018. "Spatially designed revegetation—Why the spatial arrangement of plants should be as important to revegetation as they are to natural systems." *Restoration Ecology* 26:446–55.

McCrackin, M. L., H. P. Jones, P. C. Jones, and D. Moreno-Mateos. 2017. "Recovery of lakes and coastal marine ecosystems from eutrophication: A global meta-analysis." *Limnology and Oceanography* 62:507–18.

McDonald, T., G. Gann, J. Jonson, and K. W. Dixon. 2016. *International Standards for the Practice of Ecological Restoration—Including Principles and Key Concepts.* Washington, DC: Society for Ecological Restoration.

McIntire, E. J. B., C. B. Schultz, and E. E. Crone. 2007. "Designing a network for butterfly habitat restoration: Where individuals, populations and landscapes interact." *Journal of Applied Ecology* 44:725–36.

Meli, P., M. Martínez-Ramos, J. M. Rey-Benayas, and J. Carabias. 2014. "Combining ecological, social and technical criteria to select species for forest restoration." *Applied Vegetation Science* 17:744–53.

Melis, T. S., J. Korman, and T. A. Kennedy. 2012. "Abiotic and biotic responses of the Colorado River to controlled floods at Glen Canyon Dam, Arizona, USA." *River Research and Applications* 28:764–76.

Mendenhall, C. D., C. H. Sekercioglu, F. O. Brenes, P. R. Ehrlich, and G. C. Daily. 2011. "Predictive model for sustaining biodiversity in tropical countryside." *Proceedings of the National Academy of Sciences of the United States of America* 108:16313–16.

Meretsky, V. J., D. L. Wegner, and L. E. Stevens. 2000. "Balancing endangered species and ecosystems: A case study of adaptive management in Grand Canyon." *Environmental Management* 25:579–86.

Merkel, F. R. 2010. "Evidence of recent population recovery in common eiders breeding in western Greenland." *Journal of Wildlife Management* 74:1869–74.

Metzger, J. P., and P. H. S. Brancalion. 2016. "Landscape ecology and restoration processes." In *Foundations of Restoration Ecology*, 2nd ed., edited by M. A. Palmer, J. B. Zedler, and D. A. Falk, 90–120. Washington, DC: Island Press.

Michener, W. K. 1997. "Quantatively evaluating restoration experiments: Research design, statistical analysis, and data management considerations." *Restoration Ecology* 5:324–37.

Middleton, E. L., and J. D. Bever. 2012. "Inoculation with a native soil community advances succession in a grassland restoration." *Restoration Ecology* 20:218–26.

Millar, C. I., and L. B. Brubaker. 2006. "Climate change and paleoecology: New contexts for restoration ecology." In *Foundations of Restoration Ecology*, edited by D. A. Falk, M. A. Palmer, and J. B. Zedler, 315–40. Washington, DC: Island Press.

Millenium Ecosystem Assessment. 2015. *Ecosystems and Human Well-Being: Synthesis.* Washington, DC: Island Press.

Montalvo, A., and N. Ellstrand. 2000. "Transplantation of the subshrub *Lotus scoparius*: Testing the home-site advantage hypothesis." *Conservation Biology* 14:1034–45.

Moody, M. E., and R. N. Mack. 1988. "Controlling the spread of plant invasions: The importance of nascent foci." *Journal of Applied Ecology* 25:1009–21.

Moreno-Mateos, D., E. B. Barbier, P. C. Jones, H. P. Jones, J. Aronson, J. A. López-López, M. L. McCrackin, et al. 2017. "Anthropogenic ecosystem disturbance and the recovery debt." *Nature Communications* 8: article 14163. http://dx.doi.org/10.1038/ncomms14163.

Moreno-Mateos, D., M. E. Power, F. A. Comín, and R. Yockteng. 2012. "Structural and functional loss in restored wetland ecosystems." *PLOS Biology* 10: article e1001247. https://doi.org/10.1371/journal.pbio.1001247

Morrison, M. J. 2009. *Restoring Wildlife.* Washington, DC: Island Press.

Munshower, F. F. 1994. *Practical Handbook of Disturbed Land Revegetation.* Boca Raton, FL: Lewis Publishers.

Murcia, C. 1995. "Edge effects in fragmented forests: Implications for conservation." *Trends in Ecology and Evolution* 10:58–62.

Murcia, C., and J. Aronson. 2014. "Intelligent tinkering in ecological restoration." *Restoration Ecology* 22:279–83.

Murcia, C., J. Aronson, G. H. Kattan, D. Moreno-Mateos, K. Dixon, and D. Simberloff. 2014. "A critique of the 'novel ecosystem' concept." *Trends in Ecology and Evolution* 29:548–53.

Murcia, C., M. R. Guariguata, Á. Andrade, G. I. Andrade, J. Aronson, E. M. Escobar, A. Etter, et al. 2016. "Challenges and prospects for scaling-up ecological restoration to meet international commitments: Colombia as a case study." *Conservation Letters* 9:213–20.

Narayan, S., M. W. Beck, B. G. Reguero, I. J. Losada, B. van Wesenbeeck, N. Pontee, J. N. Sanchirico, et al. 2016. "The effectiveness, costs and coastal protection benefits of natural and nature-based defences." *PLOS ONE* 11: article e0154735. https://doi.org/10.1371/journal.pone.0154735.

National Research Council. 1992. *Restoration of Aquatic Ecosystems.* Washington, DC: National Academy Press.

National Research Council. 2001. *Compensating for Wetland Losses Under the Clean Water Act.* Washington, DC: National Academy Press.

Nature Conservancy. 2018. "Insuring nature to ensure a resilient future." Accessed

March 15, 2019. https://www.nature.org/en-us/what-we-do/our-insights/perspectives/insuring-nature-to-ensure-a-resilient-future/.

Nilsson, C., T. Riis, J. M. Sarneel, and K. Svavarsdóttir. 2018. "Ecological restoration as a means of managing inland flood hazards." *BioScience* 68:89–99.

NJCWRP. n.d. "New Jersey corporate wetlands partnership." Accessed April 24, 2018. http://www.njcwrp.org/.

Norton, D. A. 2009. "Species invasions and the limits to restoration: Learning from the New Zealand experience." *Science* 325:569–71.

Ogutu-Ohwayo, R. 1990. "The decline of the native fishes of lakes Victoria and Kyoga (East Africa) and the impact of introduced species, especially the Nile perch, *Lates niloticus*, and the Nile tilapia, *Oreochromis niloticus*." *Environmental Biology of Fishes* 27:81–96.

Osenberg, C. W., B. M. Bolker, J.-S. S. White, C. M. St. Mary, and J. S. Shima. 2006. "Statistical issues and study design in ecological restorations: Lessons learned from marine reserves." In *Foundations of Restoration Ecology*, edited by D. A. Falk, M. A. Palmer, and J. B. Zedler, 280–302. Washington, DC: Island Press.

Osland, M. J., N. M. Enwright, R. H. Day, C. A. Gabler, C. L. Stagg, and J. B. Grace. 2016. "Beyond just sea-level rise: Considering macroclimatic drivers within coastal wetland vulnerability assessments to climate change." *Global Change Biology* 22:1–11.

Pagiola, S. 2008. "Payments for environmental services in Costa Rica." *Ecological Economics* 65:712–24.

Palmer, M. A., D. A. Falk, and J. B. Zedler. 2006. "Ecological theory and restoration ecology." In *Foundations of Restoration Ecology*, edited by D. A. Falk, M. A. Palmer, and J. B. Zedler, 1–10. Washington, DC: Island Press.

Palmer, M. A., K. L. Hondula, and B. J. Koch. 2014. "Ecological restoration of streams and rivers: Shifting strategies and shifting goals." *Annual Review of Ecology, Evolution, and Systematics* 45:247–69.

Palmer, M. A., and J. Ruhl. 2015. "Aligning restoration science and the law to sustain ecological infrastructure for the future." *Frontiers in Ecology and the Environment* 13:512–19.

Palmer, M. A., J. B. Zedler, and D. A. Falk (eds.). 2016. *Foundation of Restoration Ecology.* 2nd ed. Washington, DC: Island Press.

Parker, V. T., and K. E. Boyer. 2017. "Sea-level rise and climate change impacts on an urbanized Pacific Coast estuary." *Wetlands.* https://doi.org/10.1007/s13157-017-0980-7.

Pauly, D. 1995. "Anecdotes and the shifting baseline syndrome of fisheries." *Trends in Ecology and Evolution* 10:430.

Perry, D., and G. Perry. 2008. "Improving interactions between animal rights groups and conservation biologists." *Conservation Biology* 22:27–35.

Peters, M. A., D. Hamilton, and C. Eames. 2015. "Action on the ground: A review of community environmental groups, restoration objectives, activities and partnerships in New Zealand." *New Zealand Journal of Ecology* 39:179–89.

Pilon-Smits, E. A., and J. L. Freeman. 2006. "Environmental cleanup using plants: Biotechnological advances and ecological considerations." *Frontiers in Ecology and the Environment* 4:203–10.

Pimentel, D., R. Zuniga, and D. Morrison. 2005. "Update on the environmental and economic costs associated with alien-invasive species in the United States." *Ecological Economics* 52:273–88.

Pípalová, I. 2006. "A review of grass carp use for aquatic weed control and its impact on water bodies." *Journal of Aquatic Plant Management* 44:1–12.

Pistorius, T., and H. Freiberg. 2014. "From target to implementation: Perspectives for the international governance of forest landscape restoration." *Forests* 5:482–97.

Potts, M. D., T. Holland, B. F. N. Erasmus, S. Arnhold, S. Athayde, C. J. Carlson, M. S. Fennessy, et al. 2018. "Land degradation and restoration associated with changes in ecosystem services and functions, and human well-being and good quality of life." In *The IPBES Assessment Report on Land Degradation and Restoration*, edited by L. Montanarella, R. Scholes, and A. Brainich, 341–432. Bonn: Secretariat of the Intergovernmental Science-Policy Platform on Biodiversity and Ecosystem Services.

Prach, K., and R. del Moral. 2015. "Passive restoration is often quite effective: Response to Zahawi et al. (2014)." *Restoration Ecology* 23:344–46.

Pretty, J. N., C. F. Mason, D. B. Nedwell, R. E. Hine, S. Leaf, and R. Dils. 2003. "Environmental costs of freshwater eutrophication in England and Wales." *Environmental Science and Technology* 37:201.

Prober, S. M., V. A. J. Doerr, L. M. Broadhurst, K. J. Williams, and F. Dickson. 2019. "Shifting the conservation paradigm: A synthesis of options for renovating nature under climate change." *Ecological Monographs* 89: article e01333. https://esajour nals.onlinelibrary.wiley.com/doi/abs/10.1002/ecm.1333.

Raddum, G. G., A. Fjellheim, and B. L. Skjelkvåle. 2001. "Improvements in water quality and aquatic ecosystems due to reduction in sulphur deposition in Norway." *Water, Air, and Soil Pollution* 130:87–98.

Reid, J. L., M. E. Fagan, J. Lucas, J. Slaughter, and R. A. Zahawi. 2019. "The ephemerality of secondary forests in southern Costa Rica." *Conservation Letters* 12: article e12607. https://onlinelibrary.wiley.com/doi/abs/10.1111/conl.12607.

Rein, F. A., M. Los Huertos, K. D. Holl, and J. H. Langenheim. 2007. "Restoring native grasses as vegetative buffers in a coastal California agricultural landscape." *Madroño* 54:249–57.

Reitbergen-McCracken, J., S. Maginnis, and A. Sarre. 2007. *The Forest Landscape Restoration Handbook*. London: Routledge.

Rey Benayas, J. M., and J. M. Bullock. 2012. "Restoration of biodiversity and ecosystem services on agricultural land." *Ecosystems* 15:883–99.

Rey Benayas, J. M., and J. M. Bullock. 2015. "Vegetation restoration and other actions to enhance wildlife in European agricultural landscapes." In *Rewilding European Landscapes*, edited by H. M. Pereira and L. M. Navarro, 127–42. London: Springer.

Rey Benayas, J. M., J. M. Bullock, and A. C. Newton. 2008. "Creating woodland islets to reconcile ecological restoration, conservation, and agricultural land use." *Frontiers in Ecology and Environment* 6:329–36.

Rey Benayas, J. M., A. C. Newton, A. Diaz, and J. M. Bullock. 2009. "Enhancement of biodiversity and ecosystem services by ecological restoration: A meta-analysis." *Science* 325:1121–24.

Rieger, J., J. Stanley, and R. Traynor. 2014. *Project Planning and Management for Ecological Restoration*. Washington, DC: Island Press.

Riley, A. L. 2016. *Restoring Neighborhood Streams*. Washington, DC: Island Press.

Rodrigues, R. R., R. A. F. Lima, S. Gandolfi, and A. G. Nave. 2009. "On the restoration of high diversity forests: 30 years of experience in the Brazilian Atlantic Forest." *Biological Conservation* 142:1242–51.

Roni, P., and T. Beechie 2012. *Stream and Watershed Restoration: A Guide to Restoring Riverine Processes and Habitats.* Oxford: Wiley.

Rosenzweig, S. T., M. A. Carson, S. G. Baer, and J. M. Blair. 2016. "Changes in soil properties, microbial biomass, and fluxes of C and N in soil following post-agricultural grassland restoration." *Applied Soil Ecology* 100:186–94.

Rosgen, D. 1998. *Applied Stream Geomorphology.* Pagoda Spring, CO: Wildland Hydrology.

Sampaio, A. B., K. D. Holl, and A. Scariot. 2007. "Does restoration enhance regeneration of seasonal deciduous forests in pastures in central Brazil?" *Restoration Ecology* 15:462–71.

Sankaran, M., and T. M. Anderson. 2009. "Management and restoration in African savannas: Interactions and feedbacks." In *New Models for Ecosystem Dynamics and Restoration,* edited by R. J. Hobbs and K. N. Suding, 136–55. Washington, DC: Island Press.

Schlaepfer, M. A., D. F. Sax, and J. D. Olden. 2011. "The potential conservation value of non-native species." *Conservation Biology* 25:428–37.

Schlesinger, W. H., and E. S. Bernhardt. 2013. *Biogeochemistry: An Analysis of Global Change.* Amsterdam: Elsevier / Academic Press.

Schoukens, H., and A. Cliquet. 2016. "Biodiversity offsetting and restoration under the European Union Habitats Directive: Balancing between no net loss and deathbed conservation?" *Ecology and Society* 21: article 10. https://www.ecologyandsociety.org/vol21/iss4/art10/.

Scott, D. A., S. G. Baer, and J. M. Blair. 2017. "Recovery and relative influence of root, microbial, and structural properties of soil on physically sequestered carbon stocks in restored grassland." *Soil Science Society of America Journal* 81:50–60.

Seabloom, E. W., E. T. Borer, V. L. Boucher, R. S. Burton, K. L. Cottingham, L. Goldwasser, W. K. Gram, et al. 2003. "Competition, seed limitation, disturbance, and reestablishment of California native annual forbs." *Ecological Applications* 13:575–92.

Seddon, P. J., C. J. Griffiths, P. S. Soorae, and D. P. Armstrong. 2014. "Reversing defaunation: Restoring species in a changing world." *Science* 345:406–12.

Shier, D. M. 2006. "Effect of family support on the success of translocated black-tailed prairie dogs." *Conservation Biology* 20:1780–90.

Shier, D. M., and D. H. Owings. 2006. "Effects of predator training on behavior and post-release survival of captive prairie dogs (*Cynomys ludovicianus*)." *Biological Conservation* 132:126–35.

Shono, K., E. A. Cadaweng, and P. B. Durst. 2007. "Application of assisted natural regeneration to restore degraded tropical forestlands." *Restoration Ecology* 15:620–26.

Simberloff, D., and L. G. Abele. 1976. "Island biogeography and conservation practice." *Science* 191:285–86.

Sims, L., S. Tjosvold, D. Chambers, and M. Garbelotto. 2019. "Control of *Phytophthora* species in plant stock for habitat restoration through best management practices." *Plant Pathology* 68:196–204.

Society for Ecological Restoration Science and Policy Working Group (SER). 2004. *The SER Primer on Ecological Restoration.* Washington, DC: Society for Ecology Restoration International.

Southern California Wetlands Recovery Project. 2018. "Bolsa Chica wetland restoration." Accessed May 5, 2018. https://scwrp.org/projects/bolsa-chica-lowlands-restoration/.

Sprague, T. A., and H. L. Bateman. 2018. "Influence of seasonality and gestation on habitat selection by northern Mexican gartersnakes (*Thamnophis eques megalops*)." *PLOS ONE* 13: article e0191829. https://doi.org/10.1371/journal.pone.0191829.

Stahlheber, K. A., and C. M. D'Antonio. 2013. "Using livestock to manage plant composition: A meta-analysis of grazing in California Mediterranean grasslands." *Biological Conservation* 157:300–308.

Steensen, D. L., and T. A. Spreiter. 1992. "Watershed rehabilitation in Redwood National Park." In *Achieving Land Use Potential Through Reclamation: Proceedings of the Ninth Annual National Meeting of the American Society for Surface Mining and Reclamation*, 280–86. Duluth, MN: American Society for Surface Mining and Reclamation.

Strassburg, B. B. N., H. L. Beyer, R. Crouzeilles, A. Iribarrem, F. Barros, M. F. de Siqueira, A. Sánchez-Tapia, et al. 2019. "Strategic approaches to restoring ecosystems can triple conservation gains and halve costs." *Nature Ecology and Evolution* 3:62–70.

Stromberg, J. C. 2001. "Restoration of riparian vegetation in the south-western United States: Importance of flow regimes and fluvial dynamism." *Journal of Arid Environments* 49:17–34.

Suding, K. N., K. L. Gross, and G. R. Houseman. 2004. "Alternative states and positive feedbacks in restoration ecology." *Trends in Ecology and Evolution* 19:46–53.

Suding, K. N., E. Higgs, M. Palmer, J. B. Callicott, C. B. Anderson, M. Baker, J. J. Gutrich, et al. 2015. "Committing to ecological restoration." *Science* 348:638–40.

Suding, K. N., E. Spotswood, D. Chapple, E. Beller, and K. Gross. 2016. "Ecological dynamics and ecological restoration." In *Foundations of Restoration Ecology*, 2nd ed., edited by M. A. Palmer, J. B. Zedler, and D. A. Falk, 27–56. Washington, DC: Island Press.

Swan, K. D., J. M. McPherson, P. J. Seddon, and A. Moehrenschlager. 2016. "Managing marine biodiversity: The rising diversity and prevalence of marine conservation translocations." *Conservation Letters* 9:239–51.

Swenson, R. O., K. Whitener, and M. Eaton. 2003. "Restoring floods to floodplains: Riparian and floodplain restoration at the Consumnes River Preserve." In *California Riparian Systems: Processes and Floodplains Management, Ecology, and Restoration*, 224–29. Sacramento, CA: Riparian Habitat Joint Venture.

Tablado, Z., J. L. Tella, J. A. Sánchez-Zapata, and F. Hiraldo. 2010. "The paradox of the long-term positive effects of a North American crayfish on a European community of predators." *Conservation Biology* 24:1230–38.

Taylor, B. D., and R. L. Goldingay. 2012. "Restoring connectivity in landscapes fragmented by major roads: A case study using wooden poles as 'stepping stones' for gliding mammals." *Restoration Ecology* 20:671–78.

Telesetsky, A. 2017. "Eco-restoration, private landowners and overcoming the status quo bias." *Griffith Law Review* 26:248–74.

Telesetsky, A., A. Cliquet, and A. Akhtar-Khavari. 2017. *Ecological Restoration in International Environmental Law*. London: Routledge.

Temperton, V. M., A. Baasch, P. von Gillhaussen, and A. Kirmer. 2016. "Assembly theory for restoring ecosystem structure and functioning: Timing is everything?" In *Foundations of Restoration Ecology*, 2nd ed., edited by M. A. Palmer, J. B. Zedler, and D. A. Falk, 245–70. Washington, DC: Island Press.

Templeton, A. R., H. Brazeal, and J. L. Neuwald. 2011. "The transition from isolated patches to a metapopulation in the eastern collared lizard in response to prescribed fires." *Ecology* 92:1736–47.

Thaman, B., R. R. Thaman, A. Balawa, and J. Veitayaki. 2017. "The recovery of a tropical marine mollusk fishery: A transdisciplinary community-based approach in Navakavu, Fiji." *Journal of Ethnobiology* 37:494–513.

Tharme, R. E. 2003. "A global perspective on environmental flow assessment: Emerging trends in the development and application of environmental flow methodologies for rivers." *River Research and Applications* 19:397–441.

Thayer, G. W., T. A. McTigue, R. J. Salz, D. H. Merkey, F. M. Burrows, and P. F. Gayaldo. 2005. *Science-Based Restoration Monitoring of Coastal Habitats*. Vol. 2: *Tools for Monitoring Coastal Habitats*, Decision Analysis Series No. 23. Silver Spring, MD: NOAA Coastal Ocean Program.

Tompkins Conservation. n.d. "Landscape restoration." Accessed April 24, 2018. http://www.tompkinsconservation.org/landscape_restoration.htm.

Tongway, D. J., and J. A. Ludwig. 1996. "Rehabilitation of semiarid landscapes in Australia I. Restoring productive soil patches." *Restoration Ecology* 4:388–406.

Török, P., and A. Helm. 2017. "Ecological theory provides strong support for habitat restoration." *Biological Conservation* 206:85–91.

Tucker, N. I. G., and T. Simmons. 2009. "Restoring a rainforest habitat linkage in north Queensland: Donaghy's Corridor." *Ecological Management and Restoration* 10:98–112.

Turner Endangered Species Fund. n.d. Accessed April 2, 2019. https://tesf.org.

Uprety, Y., H. Asselin, Y. Bergeron, F. Doyon, and J.-F. Boucher. 2012. "Contribution of traditional knowledge to ecological restoration: Practices and applications." *Ecoscience* 19:225–37.

US Department of Agriculture (USDA). 2016. "Greater sage-grouse conservation and the sagebrush ecosystem." Accessed April 26, 2018. https://www.sagegrouseinitiative.com/report/.

US Fish and Wildlife Service (USFWS). 2012. "Welcome to the common murre restoration project." Accessed April 30, 2018. https://www.fws.gov/sfbayrefuges/Murre/.

Vandermeer, J., I. G. de la Cerda, D. Boucher, I. Perfecto, and J. Ruiz. 2000. "Hurricane disturbance and tropical tree species diversity." *Science* 290:788–91.

van Oppen, M. J. H., J. K. Oliver, H. M. Putnam, and R. D. Gates. 2015. "Building coral reef resilience through assisted evolution." *Proceedings of the National Academy of Sciences* 112:2307–13.

Van Vooren, L., B. Reubens, S. Broekx, P. D. Frenne, V. Nelissen, P. Pardon, and K. Verheyen. 2017. "Ecosystem service delivery of agri-environment measures: A synthesis for hedgerows and grass strips on arable land." *Agriculture, Ecosystems and Environment* 244:32–51.

van Wilgen, B. W., G. G. Forsyth, D. C. Le Maitre, A. Wannenburgh, J. D. F. Kotzé, E. van den Berg, and L. Henderson. 2012. "An assessment of the effectiveness of a large, national-scale invasive alien plant control strategy in South Africa." *Biological Conservation* 148:28–38.

van Wilgen, B. W., and A. Wannenburgh. 2016. "Co-facilitating invasive species control, water conservation and poverty relief: Achievements and challenges in South Africa's Working for Water programme." *Current Opinion in Environmental Sustainability* 19:7–17.

Veloz, S. D., N. Nur, L. Salas, D. Jongsomjit, J. Wood, D. Stralberg, and G. Ballard. 2013. "Modeling climate change impacts on tidal marsh birds: Restoration and

conservation planning in the face of uncertainty." *Ecosphere* 4: article 49. http://dx
.doi.org/10.1890/ES12-00341.1.

Viani, R. A. G., K. D. Holl, A. Padovezi, B. B. N. Strassburg, F. T. Farah, L. C. Garcia, R. B. Chaves, et al. 2017. "Protocol for monitoring tropical forest restoration."
Tropical Conservation Science 10: article 1940082917697265. https://doi.org/10.1177
/1940082917697265.

Vieira, D. L. M., K. D. Holl, and F. M. Peneireiro. 2009. "Agro-successional restoration as
a strategy to facilitate tropical forest recovery." *Restoration Ecology* 17:451–59.

Vitt, P., K. Havens, A. T. Kramer, D. Sollenberger, and E. Yates. 2010. "Assisted migration of plants: Changes in latitudes, changes in attitudes." *Biological Conservation*
143:18–27.

Vitule, J. R. S., C. A. Freire, D. P. Vazquez, M. A. Nuñez, and D. Simberloff. 2012. "Revisiting the potential conservation value of nonnative species." *Conservation Biology*
26:1153–55.

Vogl, K. 1980. "The ecological factors that produce perturbation-dependent ecosystems." In *The Recovery Process in Damaged Ecosystems*, edited by J. Cairns Jr., 63–94.
Ann Arbor, MI: Ann Arbor Science.

Walker, B. A., C. Dixon, P. Drobney, S. Jacobi, V. M. Hunt, A. McColpin, K. Viste-
Sparkman, et al. 2018. "The prairie reconstruction initiative database: Promoting
standardized documentation of reconstructions." *Ecological Restoration* 36:3–5.

Walker, G. B., S. L. Senecah, and S. E. Daniels. 2006. "From the forest to the river: Citizens' views of stakeholder engagement." *Human Ecology Review* 13:193–202.

Walker, L. R., and R. del Moral. 2003. *Primary Succession and Ecosystem Rehabilitation*.
Cambridge: Cambridge University Press.

Walters, C. 1986. *Adaptive Management of Renewable Resources*. New York: Macmillan.

Watson, James E. M., D. F. Shanahan, M. Di Marco, J. Allan, W. F. Laurance, E. W. Sanderson, B. Mackey, et al. 2016. "Catastrophic declines in wilderness areas undermine
global environment targets." *Current Biology* 26:2929–34.

Watts, C., and D. Thornburrow. 2009. "Where have all the weta gone? Results after
two decades of transferring a threatened New Zealand giant weta, *Deinacrida mahoenui*." *Journal of Insect Conservation* 13:287–95.

Wehi, P. M. 2009. "Indigenous ancestral sayings contribute to modern conservation
partnerships: Examples using *Phormium tenax*." *Ecological Applications* 19:267–75.

Westman, W. E. 1991. "Ecological restoration projects: Measuring their performance."
Environmental Professional 13:207–15.

Whisenant, S. G. 1999. *Repairing Damaged Wildlands: A Process-Oriented, Landscape-Scale
Approach*. Cambridge: Cambridge University Press.

White, P. S., and J. L. Walker. 1997. "Approximating nature's variation: Selecting and
using reference information in restoration ecology." *Restoration Ecology* 5:338–49.

Wickramasinghe, D. 2017. "Regreening the coast: Community-based mangrove conservation and restoration in Sri Lanka." In *Participatory Mangrove Management in
a Changing Climate: Perspectives from the Asia-Pacific*, edited by R. DasGupta and
R. Shaw, 161–71. Tokyo: Springer Japan.

Wikramanayake, E., A. Manandhar, S. Bajimaya, S. Nepal, G. Thapa, and K. Thapa.
2010. "The Terai Arc landscape: A tiger conservation success story in a human-
dominated landscape." In *Tigers of the World*, edited by R. Tilson and P. J. Nyhus,
163–73. Boston: William Andrew Publishing.

Wilson, C. W., R. E. Masters, and G. A. Buckenhofer. 1995. "Breeding bird response to pine-grassland community restoration for red-cockaded woodpeckers." *Journal of Wildlife Management* 59:56–67.

Wilson, S. D. 2015. "Managing contingency in semiarid grassland restoration through repeated planting." *Restoration Ecology* 23:385–92.

Witte, F., B. S. Msuku, J. H. Wanink, O. Seehausen, E. F. B. Katunzi, P. C. Goudswaard, and T. Goldschmidt. 2000. "Recovery of cichlid species in Lake Victoria: An examination of factors leading to differential extinction." *Reviews in Fish Biology and Fisheries* 10:233–41.

Wohl, E., S. N. Lane, and A. C. Wilcox. 2015. "The science and practice of river restoration." *Water Resources Research* 51:5974–97.

Woods, B. 1984. "Ants disperse seeds of herb species in a restored maple forest (Wisconsin)." *Restoration and Management Notes* 2:29–30.

World Bank. n.d. "Landscape approach to forest restoration and conservation." Accessed April 23, 2019. http://projects.worldbank.org/P131464/landscape-approach -forest-restoration-conservation-lafrec?lang=en&tab=overview.

Wright, E. C., and J. A. Souder. 2018. "Using applied science for effective watershed restoration and coho salmon recovery in coastal Oregon streams." *Case Studies in the Environment.* http://cse.ucpress.edu/content/early/2018/01/25/cse.2017.000489 .abstract.

Wubs, E. R. J., W. H. van der Putten, M. Bosch, and T. M. Bezemer. 2016. "Soil inoculation steers restoration of terrestrial ecosystems." *Nature Plants* 2:16107. https:// www.nature.com/articles/nplants2016107.

Zahawi, R. A., and C. K. Augspurger. 1999. "Early plant succession in abandoned pastures in Ecuador." *Biotropica* 31:540–52.

Zahawi, R. A., J. P. Dandois, K. D. Holl, D. Nadwodny, J. L. Reid, and E. C. Ellis. 2015. "Using lightweight unmanned aerial vehicles to monitor tropical forest recovery." *Biological Conservation* 186:287–95.

Case Studies and Online Resources

The following resources are available through this book's website: islandpress.org /restoration-primer.

Case Studies

The case studies provide detailed descriptions of eight restoration projects worldwide along with color maps and illustrations. They were chosen to illustrate a range of general concepts discussed in this book, and they are referenced in the text where relevant.

Asian Mangroves: Community involvement in mangrove restoration provides coastal hazard reduction and enhances human livelihoods, Indonesia and Sri Lanka—Karen D. Holl

Atlantic Forest: Diverse stakeholder participation leads to large-scale restoration successes in the Atlantic forest ecosystem, Brazil—Karen D. Holl and Pedro H. S. Brancalion

Elwha River: Removing dams to restore physical and ecological processes, Washington—Karen D. Holl and Amy E. East

Galapagos Tortoise: Complex ecological feedbacks of reintroducing giant tortoises to the Galapagos Islands, Ecuador—Karen D. Holl and J. Leighton Reid

Kissimmee River: Large-scale restoration of water flow, channel form, and ecological communities, Florida—Karen D. Holl and Joseph W. Koebel Jr.

Sacramento River: Balancing ecological and social restoration goals to restore riparian habitat along a lowland floodplain river, California—Karen D. Holl and Gregory H. Golet

Tamarix Removal: Riparian habitat restoration results in lawsuits and concern for an endangered bird in the western United States—Josephine C. Lesage and Tom L. Dudley

Younger Lagoon: Integrating teaching and research into restoring a highly degraded urban reserve in coastal California—Karen D. Holl and Elizabeth A. Howard

Questions for Reflection and Discussion

These questions ask the reader to reflect on the material presented in this book. Although the information presented in the book is as general as possible, restoration needs to be tailored to local biophysical and socioeconomic conditions. Therefore, many questions ask you to think about and apply the concepts discussed in the chapter to a specific restoration project, ideally one with which you are familiar.

Other Resources

The website gives links to videos, photos, design plans, and other resources that help visualize ecological restoration. These links will be periodically updated as new material becomes available.

About the Author

Karen D. Holl is a professor of environmental studies at the University of California, Santa Cruz, where she has taught courses and advised students on research and internships related to ecological restoration since 1996. Her research focuses on understanding how local and landscape-scale processes affect ecosystem recovery from human disturbance and using this information to restore damaged ecosystems. She conducts research primarily in rain forests in Latin America and chaparral, grassland, and riparian systems in California. She advises numerous land management and conservation organizations in California and internationally on ecological restoration. She was selected as an Aldo Leopold Leadership Fellow and as a Fellow of the California Academy of Sciences, and she was the 2017 cowinner of the Theodore Sperry Award of the Society for Ecological Restoration, which recognizes those who have made a significant and enduring contribution to advancing restoration science and practices. She has also served as chair of the Environmental Studies Department and as the faculty director of the Kenneth S. Norris Center for Natural History at the University of California, Santa Cruz.

Index